An Imaginary Tale

THE STORY OF $\sqrt{-1}$

Paul J. Nahin

PRINCETON UNIVERSITY PRESS

PRINCETON, NEW JERSEY

ISBN 0-691-02795-1

This book has been composed in Times Roman

Printed in the United States of America

DEDICATED TO MY MOTHER

KATHERINE DOROTHY MARKFELDER

AND TO THE MEMORY OF

MY FATHER

PAUL GILBERT NAHIN

(1916–1990)

A Note to the Reader

AN IMAGINARY TALE has a very strong historical component to it, but that does not mean it is a mathematical lightweight. But don't read too much into that either. It is *not* a scholarly tome meant to be read only by some mythical, elite group, such as the one described in the 1920s fable that claimed there were only twelve people in the world who really understood Einstein. $\sqrt{-1}$ has long labored under a similar false story of unfathomable mystery. The French philosophical genius of the Enlightenment, Denis Diderot, wrote of mathematicians that they "resemble those who gaze out from the tops of high mountains whose summits are lost in the clouds. Objects on the plain below have disappeared from view; they are left with only the spectacle of their own thoughts and the consciousness of the height to which they have risen and where perhaps it is not possible for everyone to follow and breathe [the thin air]." Well, in this book the air is almost always at sea-level pressure. Large chunks of this book can, in fact, be read and understood by a high school senior who has paid attention to his or her teachers in the standard fare of pre-college courses. Still, it will be most accessible to the million or so who each year complete a college course in freshman calculus. It is not a textbook, but I do think it could be profitably read by students as a supplement to the more standard presentations of mathematics. I am an electrical engineer, not a mathematician, and the style of the writing reflects the difference. Indeed, I have taken advantage of my freedom from the usual dictates of textbook pedagogy—which in its worst form can be pedantic—to write in a casual and, I hope, entertaining style. But when I need to do an integral, let me assure you I have not fallen to my knees in dumbstruck horror. And neither should you. The power and beauty of complex numbers and functions, and the amazing story of their discovery, will amply repay any extension of concentration that the relatively harder parts may demand.

Contents

Illustrations

Preface

LONG AGO, in a year so far in the past (1954) that my life then as a high school freshman now seems like a dream, my father gave me the gift of a subscription to a new magazine called *Popular Electronics*. He did this because he was a scientist, and his oldest son seemed to have talents in science and mathematics that were in danger of being subverted by the evil of science fiction. I had, in fact, given him plenty of reason for such concern. I devoured science fiction in those days, you see, often sitting in the kitchen at eleven at night eating a huge sandwich and reading a novel set on Mars a million years in the future. Dad, of course, would have preferred that I be reading a book on algebra or physics.

Being a clever man, he decided not to simply forbid the science fiction, but rather to outflank the science fiction stories by getting me to read *technical* stories, like the "Carl and Jerry" tales that appeared each month in *Popular Electronics*. Carl and Jerry were two high school electronics whiz kids— geeks or nerds, in today's unattractive terms—who managed each month to get involved in some exciting adventure in which their technical knowledge saved the day. They were a 1950s amalgamation of the Hardy Boys and Tom Swift. My father's plan was to get me to identify with Carl and Jerry, instead of with Robert Heinlein's neurotic time travelers.

Well, Dad's devious plan worked (although I never completely gave up the science fiction), and I got hooked not only on Carl and Jerry but also on the electronic construction projects the magazine featured in each issue. I learned how to read electrical schematics from the magazine, whose editors used the same exploded-view, pictorial wiring diagrams that became so well known to all who ever built an electronic, mail-order kit. I constructed a home workshop in the garage behind the house, and a lot of amazing gadgets were built there—although not all of them worked, or at least not in the way the original designers had intended.

My greatest success was an "applause meter" that was used by the judges at the high school talent show one year—a loudspeaker pick-up, an audio amplifier, and a 500 microampere meter wired into the amplifier output was all it really was. But what really had the greatest impact on me wasn't that gadget, or any of the others that I built during my high school years. It was one that, in a burst of youthful enthusiasm exceeded only by my enormous ignorance of theory, I didn't realize is *impossible* to build.

When the April 1955 issue of *Popular Electronics* arrived in the mail, one of the inside photographs displayed an incredible sight—a desk lamp emitting not a cone of light, but, instead, a cone of darkness! My eyes bugged out when I saw that. What wondrous science was at work here, I gasped (metaphorically speaking, of course, because what fourteen-year-old kid do you know, other than in a TV sitcom, who actually talks like that?). The secret, according to the accompanying article, was that the lamp was not plugged into a normal power outlet, but rather into an outlet delivering *contra-polar power.* Another photograph showed a soldering iron plugged into the contra-polar power outlet—it was covered with ice! And another displayed a frozen ice tray on a hot plate, except it was now a cold plate because it was plugged into contra-polar power. I looked at those three photographs, and I remember my pulse rate elevated and I felt a momentary spell of faintness. This was simply wonderful.

Well, of course it was all just a huge editorial joke, aided by some nifty photographic retouching. When I showed the article to my father, he glanced at it and then looked at me with what I now know was a mixture of pity and amusement. Dad wasn't an electrical engineer or a physicist, but with a Ph.D. in chemistry he wasn't totally ignorant of technical matters that fell outside the realm of benzene rings and molecular bonds. He immediately suspected that "contra-polar power" probably violated perhaps seven different fundamental principles of physics. Rather than laughing at me, however, he simply said, "Son, look at the date on the cover." I had not noticed the "April" before, or even the subtitle, "In keeping with the first day of April," but I quickly understood the significance. I still remember my enormous embarrassment at having been so completely taken in. Even I could recognize the "spoofiness" of contra-polar power once I got to the end of the article and read its footnote 4. It listed a phony citation, as follows: "Transactions of the Contra-Polar Energy Commission, Vol. 45, pp. 1324–1346 (Ed. Note—A reprint of a document found in a flying saucer)."

Like any good spoof, it had lots of tantalizing truths in it, but presented in a slightly goofy way. To give you a sample of the tone of the article, here's a typical passage: "When 'contra-polar energy' is applied to an ordinary table lamp, light is not produced, but taken away, and the area affected by the lamp becomes dark. (*Editor's Note:* This phenomenon should not be confused with 'black light,' so-called, which actually is merely light without any visible elements. As far as the human eye is concerned, 'black light' is equivalent to zero light; the light produced by contra-polar energy might be designated 'negative light,' since it subtracts from light already present.)"

To set readers up for an "explanation" of the astounding properties of contra-polar energy, the very next sentence makes the following assertion, hilarious now, as I read it decades later, but quite logical to me in 1955: "One of the reasons why atomic energy has not yet become popular among home experimenters is that an understanding of its production requires a knowledge of very advanced mathematics." Just algebra, however, would strip bare contra-polar energy, or so claimed the article. Contra-polar power "worked" by simply using the negative square root (instead of the positive root) in calculating the resonant frequency in a circuit containing both inductance and capacitance. The idea of negative frequency was intriguing to my mind (and electrical engineers have actually made sense out of it when combined with $\sqrt{-1}$), but then the editors played a few more clever tricks and came up with negative resistance.

As everybody knows who likes to curl up in a warm bed at night under an electric blanket, or who likes a crunchy piece of toast in the morning, resistors (positive resistors) get hot when conducting an electric current. "Obviously," then, a negative resistor should cool down when conducting a current—hence the soldering iron and the ice cube tray photographs. (The logic, if you can call it that, behind the cone of darkness out of a desk lamp still eludes me, however.) Now, there really is such a thing as negative resistance, and it has long been known by electrical engineers to occur in the operation, under certain specific conditions, of electric arcs. Such arcs were used, for example, in the very early pre-electronic days of radio to build extremely powerful transmitters that were able to broadcast music and human speech, rather than just the on-off telegraph code signals that were all the Hertz and Marconi spark-gap transmitters could send. Later, in college, I would learn that the operation of radio is impossible to understand, at a deep theoretical level, without an understanding of $\sqrt{-1}$.

All of this was more fascinating to my young mind than I can tell you, even forty years later, even with a somewhat increased vocabulary. It showed me that there were big, exciting ideas in the world of electronics, bigger than I had ever imagined while out in the garage tinkering with my gadgets. And later, when in my high school algebra classes I was introduced to complex numbers as the solutions to certain quadratic equations, I knew (unlike my mostly perplexed classmates) that they were not just part of a sterile intellectual game. I already knew that $\sqrt{-1}$ was important to electrical engineers, and to their ability to construct truly amazing devices.

Three and a half years after reading about contra-polar power, I was sitting in an early morning train out of Los Angeles' Union Station, heading north to Palo Alto on my way to join the Stanford University freshman class of 1958. Over the years that Carl and Jerry appeared in *Popular Electronics,* the tales chronicled their progression from high school freshmen to electrical engineering students at the fictional "Parvoo University" and, like them, I was taking the first step on the career path I've trodden ever since, as an electrical engineer. Once at Stanford I had more than enough to fill my days with reading and so I quickly drifted away from *Popular Electronics,* but it had been there at just the right time for me; Dad's plan had worked better than he could have possibly hoped. In a certain sense, then, my whole professional life has been the result of my youthful fascination with the mystery of $\sqrt{-1}$ and that is why I have written this book.[1]

In a letter (dated January 13, 1852) to his English friend Augustus De Morgan, the Irish mathematician William Rowan Hamilton wrote, "I see that either *you or I*—but I hope it will be you—must write, some time or other, a history of $\sqrt{-1}$." Five days later De Morgan replied, "As to a history of $\sqrt{-1}$, it would be no small job to do it well from the Hindoos downwards." Well, neither Hamilton or De Morgan ever wrote that history and, as far as I know, nobody else has either. And so that's another reason why I wrote this book. I simply wanted to learn more.

My one great regret is that Dad isn't here to read it. But if he were, I hope he would be pleased at the result of his investment in a magazine subscription nearly a half century ago.

An Imaginary Tale

Introduction

IN 1878 a pair of brothers, the soon-to-become-infamous thieves Ahmed and Mohammed Abd er-Rassul, stumbled upon the ancient Egyptian burial site in the Valley of Kings, at Deir el-Bahri. They quickly had a thriving business going selling stolen relics, one of which was a mathematical papyrus; one of the brothers sold it to the Russian Egyptologist V. S. Golenishchev in 1893, who in turn gave it to the Museum of Fine Arts in Moscow in 1912.[1] There it remained, a mystery until its complete translation in 1930, at which time the scholarly world learned just how mathematically advanced the ancient Egyptians had been.

In particular, the fourteenth problem of the Moscow Mathematical Papyrus (MMP), as it is now called, is a specific numerical example of how to calculate the volume V of a truncated square pyramid, the so-called *frustum* of a pyramid. This example strongly suggests that the ancient Egyptians knew the formula

$$V = \frac{1}{3}h(a^2 + ab + b^2),$$

where a and b are the edge lengths of the bottom and top squares, respectively, and h is the height. One historian of science has called this knowledge "breath-taking" and "the masterpiece of Egyptian geometry."[2] The derivation of this formula is a routine exercise for anyone who has had freshman calculus, but it is much less obvious how the Egyptians could have discovered it without a knowledge of integral calculus.[3]

While correct, this result does have one very slight stylistic flaw. The values of a and b are what a modern engineer or physicist would call an "observable," i.e., they are lengths that can be directly determined simply by laying a tape measure out along the bottom and top edges of the frustum. The value of h, however, is not directly measurable, or certainly it isn't for a solid pyramid. It can be calculated for any given pyramid, of course, using a knowledge of geometry and trigonometry, but how much more direct it would be to express the volume of the frustum in terms not of h, but of c, the *slant* edge length. That length *is* directly measurable. This was finally done but, as far as is known today, not until the first century A.D. by the great mathematician-engineer Heron of Alexandria, who is usually called a Greek but may have actually been an Egyptian. It is, in fact, an elementary problem in geometry to show that

$$h = \sqrt{c^2 - 2\left(\frac{a-b}{2}\right)^2}.$$

Now, let's skip ahead in time to 1897, to a talk given that year at a meeting of the American Association for the Advancement of Science by Wooster Woodruff Beman, a professor of mathematics at the University of Michigan, and a well-known historian of the subject. I quote from that address:

> We find the square root of a negative quantity appearing for the first time in the *Stereometria* of Heron of Alexandria . . . After having given a correct formula for the determination of the volume of a frustum of a pyramid with square base and applied it successfully to the case where the side of the lower base is 10, of the upper 2, and the edge 9, the author endeavors to solve the problem where the side of the lower base is 28, of the upper 4, and the edge 15. Instead of the square root of $81 - 144$ required by the formula, he takes the square root of $144 - 81$ . . . , i.e., he replaces $\sqrt{-1}$ by 1, and fails to observe that the problem as stated is impossible. Whether this mistake was due to Heron or to the ignorance of some copyist cannot be determined.[4]

That is, using $a = 28$, $b = 4$, and $c = 15$ in his formula for h, Heron wrote:

$$h = \sqrt{(15)^2 - 2\left(\frac{28-4}{2}\right)^2} = \sqrt{225 - 2(12)^2} = \sqrt{225 - 144 - 144} = \sqrt{81 - 144}.$$

The next, magnificent step would of course have been to write $h = \sqrt{-63}$, but the *Stereometria* records it as $h = \sqrt{63}$, and so Heron missed being the earliest known scholar to have derived the square root of a negative number in a mathematical analysis of a physical problem. If Heron really did fudge his arithmetic then he paid dearly for it in lost fame. It would be a thousand years more before a mathematician would even bother to take notice of such a thing—and then simply to dismiss it as obvious nonsense—and yet five hundred years more before the square root of a negative number would be taken seriously (but still be considered a mystery).

While Heron almost surely had to be aware of the appearance of the square root of a negative number in the frustum problem, his fellow Alexandrian two centuries later, Diophantus, seems to have completely missed a similar event when he chanced upon it. Diophantus is honored today as having played the same role in algebra that Euclid did in geometry. Euclid gave us his *Elements,* and Diophantus presented posterity with the *Arithmetica.* In both of these cases, the information contained was almost certainly the results of many previous, anonymous mathematicians whose identities are now lost forever to history. It was Euclid and Diophantus, however, who collected and organized this mathematical heritage in coherent form in their great works.

In my opinion, Euclid did the better job because *Elements* is a logical *theory* of plane geometry. *Arithmetica,* or at least the several chapters or books that have survived of the original thirteen, is, on the other hand, a collection of specific numerical solutions to certain problems, with no generalized, theoretical development of methods. Each problem in *Arithmetica* is unique unto itself, much like those on the Moscow Mathematical Papyrus. But this is not to say that the solutions given are not ingenious, and in many cases even diabolically clever. *Arithmetica* is still an excellent hunting ground for a modern teacher of high school algebra looking for problems to challenge, even stump, the brightest of students.[5]

In book 6, for example, we find the following problem (number 22): Given a right triangle with area 7 and perimeter 12, find its sides. Here's how Diophantus derived the quadratic equation $172x = 336x^2 + 24$ from the statement of the problem. With the sides of the right triangle denoted by P_1 and P_2, the problem presented by Diophantus is equivalent to solving the simultaneous equations

$$P_1 P_2 = 14,$$

$$P_1 + P_2 + \sqrt{P_1^2 + P_2^2} = 12.$$

These can be solved by routine, if somewhat lengthy, algebraic manipulation, but Diophantus' clever idea was to immediately reduce the number of variables from two to one by writing

$$P_1 = \frac{1}{x} \text{ and } P_2 = 14x.$$

Then the first equation reduces to the identity $14 = 14$, and the second to

$$\frac{1}{x} + 14x + \sqrt{\frac{1}{x^2} + 196x^2} = 12,$$

which is easily put into the form given above,

$$172x = 336x^2 + 24.$$

It is a useful exercise to directly solve the original P_1, P_2 equations, and then to show that the results are consistent with Diophantus' solution.

Diophantus wrote the equation the way he did because it displays all the coefficients as positive numbers, i.e., the ancients rejected negative numbers as being without meaning because they could see no way physically to interpret a number that is "less than nothing." Indeed, elsewhere in *Arithmetica* (problem 2 in book 5) he wrote, of the equation $4x + 20 = 4$, that it was "absurd" because it would lead to the "impossible" solution $x = -4$. In accordance with this position, Diophantus used only the positive root when solving a quadratic. As late as the sixteenth century we find mathematicians referring to the negative roots of an equation as *fictitious* or *absurd* or *false*.

And so, of course, the square root of a negative number would have simply been beyond the pale. It is the French mathematician René Descartes, writing fourteen centuries later in his 1637 *La Geometrie,* whose work I will discuss in some detail in chapter 2, to whom we owe the term *imaginary* for such numbers. Before Descartes' introduction of this term, the square roots of negative numbers were called *sophisticated* or *subtle*. It is just such a thing, in fact, that Diophantus' quadratic equation for the triangle problem results in, i.e., the quadratic formula quickly gives the solutions

$$x = \frac{43 \pm \sqrt{-167}}{168}.$$

But this is not what Diophantus wrote. What he wrote was simply that the quadratic equation was not possible. By that he meant the equation has no rational solution because "half the coefficient of x multiplied into itself, minus the product of the coefficient of x^2 and the units" must make a square for a rational solution to exist, while

$$\left(\frac{172}{2}\right)^2 - (336)(24) = -668$$

certainly is not a square. As for the square root of this negative number, Diophantus had nothing at all to say.

Six hundred years later (circa 850 A.D.) the Hindu mathematician Mahaviracarya wrote on this issue, but then only to declare what Heron and Diophantus had practiced so long before: "The square of a positive as well as of a negative (quantity) is positive; and the square roots of those (square quantities) are positive and negative in order. As in the nature of things a negative (quantity) is not a square (quantity), *it has therefore no square root* [my emphasis]."[6] More centuries would pass before opinion would change.

At the beginning of George Gamow's beautiful little book of popularized science, *One Two Three . . . Infinity,* there's the following limerick to give the reader a flavor both of what is coming next, and of the author's playful sense of humor:

There was a young fellow from Trinity
Who took $\sqrt{\infty}$.
But the number of digits
Gave him the fidgets;
He dropped Math and took up Divinity.

This book is not about the truly monumental task of taking the square root of infinity, but rather about another task that a great many very clever mathematicians of the past (certainly including Heron and Diophantus) thought an even more absurd one—that of figuring out the meaning of the square root of minus one.

The Puzzles of Imaginary Numbers

1.1 THE CUBIC EQUATION

At the end of his 1494 book *Summa de Arithmetica, Geometria, Proportioni et Proportionalita,* summarizing all the knowledge of that time on arithmetic, algebra (including quadratic equations), and trigonometry, the Franciscan friar Luca Pacioli (circa 1445–1514) made a bold assertion. He declared that the solution of the cubic equation is "as impossible at the present state of science as the quadrature of the circle." The latter problem had been around in mathematics ever since the time of the Greek mathematician Hippocrates, circa 440 B.C. The quadrature of a circle, the construction by straightedge and compass alone of the square equal in area to the circle, had proven to be difficult, and when Pacioli wrote the quadrature problem was still unsolved. He clearly meant only to use it as a measure of the difficulty of solving the cubic, but actually the quadrature problem is a measure of the *greatest* difficulty, since it was shown in 1882 to be impossible.

Pacioli was wrong in his assertion, however, because within the next ten years the University of Bologna mathematician Scipione del Ferro (1465–1526) did, in fact, discover how to solve the so-called *depressed cubic,* a special case of the general cubic in which the second-degree term is missing. Because his solution to the depressed cubic is central to the first progress made toward understanding the square root of minus one, it is worth some effort in understanding just what del Ferro did.

The general cubic contains all the powers of the unknown, i.e.,

$$x^3 + a_1x^2 + a_2x + a_3 = 0$$

where we can take the coefficient of the third-degree term to be unity without loss of any generality. If that coefficient is not one, then we just divide through the equation by the coefficient, which we can always do unless it is zero—but then the equation isn't really a cubic.

The cubic solved by del Ferro, on the other hand, has the general form of

$$x^3 + px = q,$$

where p and q are non-negative. Just like Diophantus, sixteenth-century mathematicians, del Ferro included, avoided the appearance of negative coefficients in their equations.[1] Solving this equation may seem to fall somewhat short of solving the general cubic, but with the discovery of one last ingenious trick del Ferro's solution is general. What del Ferro somehow stumbled upon is that solutions to the depressed cubic can be written as the sum of two terms, i.e., we can express the unknown x as $x = u + v$. Substituting this into the depressed cubic, expanding, and collecting terms, results in

$$u^3 + v^3 + (3uv + p)(u + v) = q.$$

This single, rather complicated-looking equation, can be rewritten as two individually less complicated statements:

$$3uv + p = 0$$

which then says

$$u^3 + v^3 = q.$$

How did del Ferro know to do this? The Polish-American mathematician Mark Kac (1914–84) answered this question with his famous distinction between the ordinary genius and the magician genius: "An ordinary genius is a fellow that you and I would be just as good as, if we were only many times better. There is no mystery as to how his mind works. Once we understand what he has done, we feel certain that we, too, could have done it. It is different with the magicians . . . the working of their minds is for all intents and purposes incomprehensible. Even after we understand what they have done, the process by which they have done it is completely dark." Del Ferro's idea was of the magician class.

Solving the first equation for v in terms of p and u, and substituting into the second equation, we obtain

$$u^6 - qu^3 - \frac{p^3}{27} = 0.$$

At first glance this sixth-degree equation may look like a huge step backward, but in fact it isn't. The equation is, indeed, of the sixth degree, but it is also quadratic in u^3. So, using the solution formula for quadratics, well-known since Babylonian times, we have

$$u^3 = \frac{q}{2} \pm \sqrt{\frac{q^2}{4} + \frac{p^3}{27}}.$$

or, using just the positive root,[2]

9

$$u = \sqrt[3]{\frac{q}{2} + \sqrt{\frac{q^2}{4} + \frac{p^3}{27}}}.$$

Now, since $v^3 = q - u^3$, then

$$v = \sqrt[3]{\frac{q}{2} - \sqrt{\frac{q^2}{4} + \frac{p^3}{27}}}.$$

Thus, a solution to the depressed cubic $x^3 + px = q$ is the fearsome-looking expression

$$x = \sqrt[3]{\frac{q}{2} + \sqrt{\frac{q^2}{4} + \frac{p^3}{27}}} + \sqrt[3]{\frac{q}{2} - \sqrt{\frac{q^2}{4} + \frac{p^3}{27}}}.$$

Alternatively, since $\sqrt[3]{-1} = -1$, then the second term in this expression can have a -1 factor taken through the outer radical to give the equivalent

$$x = \sqrt[3]{\frac{q}{2} + \sqrt{\frac{q^2}{4} + \frac{p^3}{27}}} - \sqrt[3]{-\frac{q}{2} + \sqrt{\frac{q^2}{4} + \frac{p^3}{27}}}.$$

You can find both forms stated in different books discussing cubics, but there is no reason to prefer one over the other.

Since p and q were taken by del Ferro to be positive, it is immediately obvious that these two (equivalent) expressions for x will always give a real result. In fact, although there are three solutions or *roots* to any cubic (see appendix A), it is not hard to show that there is always exactly one real, positive root and therefore two complex roots to del Ferro's cubic (see box 1.1).

Now, before continuing with the cubic let me say just a bit about the nature of complex numbers. A *complex* number is neither purely real nor purely imaginary, but rather is a composite of the two. That is, if a and b are both purely real, then $a + b\sqrt{-1}$ is complex. The form used by mathematicians and nearly everybody else is $a + ib$ (the great eighteenth-century Swiss mathematician Leonhard Euler, about whom much more is said in chapter 6, introduced the i symbol for $\sqrt{-1}$ in 1777). This is written as $a + jb$ by electrical engineers. The reason electrical engineers generally opt for j is that $\sqrt{-1}$ often occurs in their problems when electric currents are involved, and the letter i is traditionally reserved for that quantity. Contrary to popular myth, however, I can assure you that most electrical engineers are *not* confused when they see an equation involving complex numbers written with $i = \sqrt{-1}$ rather than with j. With that said, however, let me admit that in chapter 5 I,

BOX 1.1

THE ONE REAL, POSITIVE SOLUTION TO DEL FERRO'S CUBIC EQUATION

To see that there is exactly one real, positive root to the depressed cubic $x^3 + px = q$, where p and q are both non-negative, consider the function

$$f(x) = x^3 + px - q.$$

Del Ferro's problem is that of solving for the roots of $f(x) = 0$. Now, if you calculate the derivative of $f(x)$ [denoted by $f'(x)$] and recall that the derivative is the slope of the curve $f(x)$, then you will get

$$f'(x) = 3x^2 + p,$$

which is always non-negative because x^2 is never negative, and we are assuming p is non-negative. That is, $f(x)$ always has non-negative slope, and so never decreases as x increases. Since $f(0) = -q$, which is never positive (since we are assuming that q is non-negative), then a plot of $f(x)$ must look like figure 1.1. From the figure it is clear that the curve crosses the x-axis only once, thus locating the real root, and that the crossing is such that the root is never negative (it is zero only if $q = 0$).

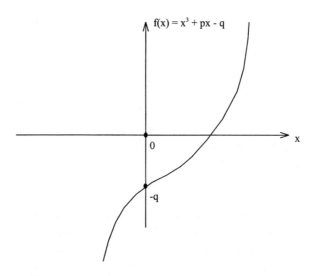

$f(x) = x^3 + px - q$

Figure 1.1. Plot of $f(x) = x^3 + px - q$, p and $q \geq 0$.

11

too, will use j rather than i for $\sqrt{-1}$ when I show you a nice little electrical puzzle from the nineteenth century.

Complex numbers obey many of the obvious rules, e.g., $(a + ib)(c + id) = ac + iad + ibc + i^2bd = ac - bd + i(ad + bc)$. But you do have to be careful. For example, if a and b can both only be positive, then $\sqrt{ab} = \sqrt{a}\sqrt{b}$. But if we allow negative numbers, too, this rule fails, e.g., $\sqrt{(-4)(-9)} = \sqrt{36} = 6 \neq \sqrt{-4}\sqrt{-9} = (2i)(3i) = 6i^2 = -6$. Euler was confused on this very point in his 1770 *Algebra*.

One final, very important comment on the reals versus the complex. Complex numbers fail to have the ordering property of the reals. *Ordering* means that we can write statements like $x > 0$ or $x < 0$. Indeed, if x and y are both real, and if $x > 0$ and $y > 0$, then their product $xy > 0$. If we try to impose this behavior on complex numbers, however, then we get into trouble. An easy way to see this is by a counterexample. That is, let us suppose we *can* order the complex numbers. Then, in particular, it must be true that either $i > 0$ or $i < 0$. Suppose $i > 0$. Then, $-1 = i \cdot i > 0$, which is clearly false. So we must suppose $i < 0$, which when we multiply through by -1 (which flips the sense of the inequality) says $-i > 0$. Then, $-1 = (-i) \cdot (-i) > 0$, just as before, and it is still clearly false. The conclusion is that the original assumption of ordering leads us into contradiction, and so that assumption must be false. Now back to cubics.

Once we have the real root to del Ferro's cubic, then finding the two complex roots is not difficult. Suppose we denote the real root given by del Ferro's equation by r_1. Then we can factor the cubic as

$$(x - r_1)(x - r_2)(x - r_3) = 0 = (x - r_1)[x^2 - x(r_2 + r_3) + r_2r_3].$$

To find the two additional roots, r_2 and r_3, we can then apply the quadratic formula to

$$x^2 - x(r_2 + r_3) + r_2r_3 = 0.$$

For example, consider the case of $x^3 + 6x = 20$, where we have $p = 6$ and $q = 20$. Substituting these values into the second version of del Ferro's formula gives

$$x = \sqrt[3]{10 + \sqrt{108}} - \sqrt[3]{-10 + \sqrt{108}}.$$

Now, if you look at the original cubic long enough, perhaps you'll have the lucky thought that $x = 2$ works ($8 + 12 = 20$). So could that complicated-looking thing with all the radical signs that I just wrote *actually be* 2? Well, yes, it is. Run it through a hand calculator and you will see that

$$x = \sqrt[3]{20.392305} - \sqrt[3]{0.392305} = 2.7320508 - 0.7320508 = 2.$$

Then, to find the other two roots to $f(x) = 0 = x^3 + 6x - 20$, we use the fact that one factor of $f(x)$ is $(x - 2)$ to find, with some long division, that

$$(x - 2)(x^2 + 2x + 10) = x^3 + 6x - 20.$$

Applying the quadratic formula to the quadratic factor quickly gives the two complex roots (solutions to the original cubic) of

$$r_2 = -1 + 3\sqrt{-1}$$

and

$$r_3 = -1 - 3\sqrt{-1}.$$

1.2 NEGATIVE ATTITUDES ABOUT NEGATIVE NUMBERS

But this is all getting ahead of the story. Del Ferro and his fellow mathematicians did not, in fact, do any of the above sort of factoring to get the complex roots—the finding of a single, real, positive number for the solution of a cubic was all they were after. And, as long as mathematicians concerned themselves with del Ferro's original depressed cubic, then a single, real, positive root is all there is, and all was well. But what of such a cubic as $x^3 - 6x = 20$, where now we have $p = -6 < 0$? Del Ferro would never have written such a cubic, of course, with its negative coefficient, but rather would have written $x^3 = 6x + 20$ and would have considered this an entirely new problem. That is, he would have started over from the beginning to solve

$$x^3 = px + q$$

with, again, both p and q non-negative. This is totally unnecessary, however, as at no place in the solution to $x^3 + px = q$ did he ever actually use the non-negativity of p and q. That is, such assumptions have no importance, and were explicitly made simply because of an unwarranted aversion by early mathematicians to negative numbers.

This suspicion of negative numbers seems so odd to scientists and engineers today, however, simply because they are used to them and have forgotten the turmoil they went through in their grade-school years. In fact, intelligent, nontechnical adults continue to experience this turmoil, as illustrated in the following wonderful couplet, often attributed to the poet W. H. Auden:

Minus times minus is plus.
The reason for this we need not discuss.

The great English mathematician John Wallis (1616–1703), for example, whom you will meet in more detail later in the next chapter as the individual who made the first rational attempt to attach physical significance to $\sqrt{-1}$, also made some incredible assertions concerning negative numbers. In his 1665 book *Arithmetica Infinitorum,* an influential book read with great interest by the young Isaac Newton, Wallis made the following argument. Since $a \div 0$, with $a > 0$, is positive infinity, and since $a \div b$, with $b < 0$, is a negative number, then this negative number must be *greater* than positive infinity because the denominator in the second case is less than the denominator in the first case (i.e., $b < 0$). This left Wallis with the astounding conclusion that a negative number is simultaneously both less than zero and greater than positive infinity, and so who can blame him for being wary of negative numbers? And, of course, he was not alone. Indeed, the great Euler himself thought Auden's concern sufficiently meritorious that he included a somewhat dubious "explanation" for why "minus times minus is plus" in his famous textbook *Algebra* (1770).

We are bolder today. Now we simply say, okay, p is negative (so what?) and plug right into the original del Ferro formula. That is, replacing the negative p with $-p$ (where now p itself is non-negative) we have

$$x = \sqrt[3]{\frac{q}{2} + \sqrt{\frac{q^2}{4} - \frac{p^3}{27}}} - \sqrt[3]{-\frac{q}{2} + \sqrt{\frac{q^2}{4} - \frac{p^3}{27}}}$$

as the solution to $x^3 = px + q$, with p and q both non-negative. In particular, the formula tells us that the solution to $x^3 = 6x + 20$ is

$$x = \sqrt[3]{10 + \sqrt{92}} - \sqrt[3]{-10 + \sqrt{92}} = 3.4377073$$

which is indeed a solution to the cubic, as can be easily verified with a hand-held calculator.

1.3 A RASH CHALLENGE

The story of the cubic now takes a tortured, twisted path. As was the tradition in those days, del Ferro kept his solution secret. He did this because, unlike today's academic mathematicians who make their living publishing their results to earn first appointment to a junior professorship and later promotion and tenure, del Ferro and his colleagues were more like independently employed businessmen. They earned their livelihoods by challenging each other to public contests of problem solving, and the winner took all—prize money, maybe, certainly "glory," and with luck the support of an admiring and rich patron. One's chances of winning such contests were obviously enhanced by knowing how to solve problems that others could not, so secrecy was the style of the day.

In fact, del Ferro almost took the secret of how to solve depressed cubics to the grave, telling at most only a small number of close friends. As he lay dying he told one more, his student Antonio Maria Fior. While Fior was not a particularly good mathematician, such knowledge was a formidable weapon and so, in 1535, he challenged a far better known and infinitely more able mathematician, Niccolo Fontana (1500–77). Fontana had come to Fior's attention because Fontana had recently announced that he could solve cubics of the general form $x^3 + px^2 = q$. Fior thought Fontana was bluffing, that he actually had no such solution, and so Fior saw him as the perfect victim, ripe for the plucking of a public contest.

Fontana, who is better known today as simply Tartaglia ("the stammerer," because of a speech impediment caused by a terrible sword wound to the jaw that he received from an invading French soldier when he was twelve), suspected Fior had received the secret of the depressed cubic from del Ferro. Fearing it would be such cubics he would be challenged with, and not knowing how to solve them, Tartaglia threw himself with a tremendous effort into solving the depressed cubic; just before the contest day he succeeded in rediscovering del Ferro's solution of $x^3 + px = q$ for himself. This is an interesting example of how, once a problem is *known* to have a solution, others quickly find it, too—a phenomenon related, I think, to sports records, e.g., within months of Roger Bannister breaking the four-minute mile it seemed as though every good runner in the world started doing it. In any case, Tartaglia's discovery, combined with his ability to really solve $x^3 + px^2 = q$ (he had *not* been bluffing), allowed him to utterly defeat Fior. Each proposed thirty problems for the other, and while Fior could solve none of Tartaglia's, Tartaglia solved all of Fior's.

1.4 The Secret Spreads

All of this is pretty bizarre, but the story gets even better. Like del Ferro, Tartaglia kept his newly won knowledge to himself, both for the reasons I mentioned before and because Tartaglia planned to publish the solutions to both

15

types of cubics himself, in a book he thought he might one day write (but never did). When the news of his rout of Fior spread, however, it quickly reached the ear of Girolamo Cardano (1501–76), otherwise known simply as Cardan. Unlike Fior, Cardan was an outstanding intellect who, among his many talents, was an extremely good mathematician.[3] Cardan's intellectual curiosity was fired by the knowledge that Tartaglia knew the secret to the depressed cubic, and he begged Tartaglia to reveal it. After initially refusing, Tartaglia eventually yielded and told Cardan the rule, but not the derivation, for calculating solutions—and even then only after extracting a vow of secrecy.

Cardan was not a saint, but he also was not a scoundrel. He almost certainly had every intention of honoring his oath of silence, but then he began to hear that Tartaglia was not the first to solve the depressed cubic. And once he had actually seen the surviving papers of del Ferro, Cardan no longer felt bound to keep his silence. Cardan rediscovered Tartaglia's solution for himself and then published it in his book *Ars Magna* (*The Great Art*—of algebra, as opposed to the lesser art of arithmetic) in 1545. In this book he gave Tartaglia *and* del Ferro specific credit, but still Tartaglia felt wronged and he launched a blizzard of claims charging Cardan with plagiarism and worse.[4] This part of the story I will not pursue here, as it has nothing to do with $\sqrt{-1}$, except to say that Tartaglia's fear of lost fame in fact came to pass. Even though he and del Ferro indeed had priority as the true, independent discoverers of the solution to the depressed cubic, ever since *Ars Magna,* it has been known as the "Cardan formula."

Cardan was not an intellectual thief (plagiarists don't give attributions), and in fact he showed how to extend the solution of the depressed cubic to *all* cubics. This was a major achievement in itself, and it is all Cardan's. The idea is as inspired as was del Ferro's original breakthrough. Cardan started with the general cubic

$$x^3 + a_1x^2 + a_2x + a_3 = 0$$

and then changed variable to $x = y - \frac{1}{3}a_1$. Substituting this back into the general cubic, expanding, and collecting terms, he obtained

$$y^3 + \left(a_2 - \frac{1}{3}a_1^2\right)y = -\frac{2}{27}a_1^3 + \frac{1}{3}a_2a_1 - a_3.$$

That is, he obtained the depressed cubic $y^3 + py = q$ with

$$p = a_2 - \frac{1}{3}a_1^2,$$

$$q = -\frac{2}{27}a_1^3 + \frac{1}{3}a_1a_2 - a_3.$$

The depressed cubic so obtained can now be solved with the Cardan formula. For example, if you start with $x^3 - 15x^2 + 81x - 175 = 0$ and then make Cardan's change of variable $x = y + 5$, you will get

$$p = 81 - \frac{1}{3}(15)^2 = 6,$$

$$q = -\frac{2}{27}(-15)^3 + \frac{1}{3}(81)(-15) - (-175) = 20,$$

and so $y^3 + 6y = 20$. I solved this equation earlier in this chapter, getting $y = 2$. Thus, $x = 7$ is the solution to the above cubic, as a hand calculation will quickly confirm.

So it looks as if the cubic equation problem has finally been put to rest, and all is well. Not so, however, and Cardan knew it. Recall the solution to $x^3 = px + q$,

$$x = \sqrt[3]{\frac{q}{2} + \sqrt{\frac{q^2}{4} - \frac{p^3}{27}}} - \sqrt[3]{-\frac{q}{2} + \sqrt{\frac{q^2}{4} - \frac{p^3}{27}}}.$$

There is a dragon lurking in *this* version of the Cardan formula! If $q^2/4 - p^3/27 < 0$ then the formula involves the square root of a negative number, and the great puzzle was not the imaginary number itself, but something quite different. The fact that Cardan had no fear of imaginaries themselves is quite clear from the famous problem he gives in *Ars Magna,* that of dividing ten into two parts whose product is forty. He calls this problem "manifestly impossible" because it leads immediately to the quadratic equation $x^2 - 10x + 40 = 0$, where x and $10 - x$ are the two parts, an equation with the complex roots—which Cardan called *sophistic* because he could see no physical meaning to them—of $5 + \sqrt{-15}$ and $5 - \sqrt{-15}$. Their sum is obviously ten because the imaginary parts cancel, but what of their product? Cardan boldly wrote "nevertheless we will operate" and formally calculated

$$(5 + \sqrt{-15})(5 - \sqrt{-15}) = (5)(5) - (5)(\sqrt{-15}) + (5)(\sqrt{-15})$$
$$+ (\sqrt{-15})(\sqrt{-15}) = 25 + 15 = 40.$$

As Cardan said of this calculation, "Putting aside the mental tortures involved" in doing this, i.e., in manipulating $\sqrt{-15}$ just like any other number, everything works out. Still, while not afraid of such numbers, it is clear from his next words that he viewed them with more than a little suspicion: "So progresses arithmetic subtlety the end of which, as is said, is as refined as it is useless." But what *really* perplexed Cardan was the case of such square roots of negative numbers occurring in the Cardan formula for cubic equations that clearly had only real solutions.

17

1.5 How Complex Numbers Can Represent Real Solutions

To see what I mean by this, consider the problem treated by Cardan's follower, the Italian engineer-architect Rafael Bombelli (1526–72). Bombelli's fame among his contemporaries was as a practical man who knew how to drain swampy marshes, but today his fame is as an expert in algebra who explained what is really going in Cardan's formula. In his *Algebra* of 1572, Bombelli presents the cubic $x^3 = 15x + 4$ and, with perhaps just a little pondering, you can see that $x = 4$ is a solution. Then, using long division/factoring, you can easily show that the two other solutions are $x = -2 \pm \sqrt{3}$. That is, *all three* solutions are real. But look at what the Cardan formula gives, with $p = 15$ and $q = 4$. Since $q^2/4 = 4$ and $p^3/27 = 125$, then

$$x = \sqrt[3]{2 + \sqrt{-121}} - \sqrt[3]{-2 + \sqrt{-121}} = \sqrt[3]{2 + \sqrt{-121}} + \sqrt[3]{2 - \sqrt{-121}}.$$

Cardan's formula gives a solution that is the sum of the cube roots of two complex conjugates (if this word is strange to you, then you should read appendix A), and you might think that if anything isn't real it will be something as "complex" as that, right? Wrong. Cardan did not realize this; with obvious frustration he called the cubics in which such a strange result occurred "irreducible" and pursued the matter no more. It is instructive, before going further, to see why he used the term "irreducible."

Cardan was completely mystified by how to actually calculate the cube root of a complex number. To see the circular loop in algebra that caused his confusion, consider Bombelli's cubic. Let us suppose that, whatever the cube root in the solution given by the Cardan formula is, we can at least write it most generally as a complex number. For example, let us write

$$\sqrt[3]{2 + \sqrt{-121}} = u + \sqrt{-v}.$$

We wish to find both u and v (where $v > 0$). Cubing both sides gives

$$2 + \sqrt{-121} = u^3 + 3u^2\sqrt{-v} - 3uv - v\sqrt{-v}.$$

Equating the real and imaginary parts on both sides, we then get

$$u^3 - 3uv = 2,$$
$$3u^2\sqrt{-v} - v\sqrt{-v} = \sqrt{-121}.$$

Squaring both of these equations gives us another pair:

$$u^6 - 6u^4v + 9u^2v^2 = 4,$$
$$-9u^4v + 6u^2v^2 - v^3 = -121.$$

and subtracting the second equation from the first equation results in

$$u^6 + 3u^4v + 3u^2v^2 + v^3 = 125.$$

Both sides of this are perfect cubes, i.e., taking cube roots gives $u^2 + v = 5$, or $v = 5 - u^2$. Substituting this back into the $u^3 - 3uv = 2$ equation above results in $4u^3 = 15u + 2$, another cubic equation in a single variable. And, in fact, dividing through by 4 to put it in the form $u^3 = pu + q$, we have $p = \frac{15}{4}$ and $q = \frac{1}{2}$ and so, using the formula at the end of section 1.2,

$$\frac{q^2}{4} - \frac{p^3}{27} = \frac{1}{16} - \frac{3,375}{(27)(64)},$$

which is clearly negative.

That is, $4u^3 = 15u + 2$ is an irreducible cubic and will, when "solved" by the Cardan formula, result in having to calculate the cube roots of complex numbers. So we are right back where we started, faced with the problem of how to calculate such a thing. The problem seems to be stuck in a loop. No wonder Cardan called this situation "irreducible." Later, in chapter 3, you will see how mathematicians eventually discovered how to calculate *any* root of a complex number.

It was Bombelli's great insight to see that the weird expression that Cardan's formula gives for x is real, but expressed in a very unfamiliar manner (see box 1.2 for what is going on geometrically in irreducible cubics). This insight did not come easily. As Bombelli wrote in his *Algebra*, "It was a wild thought in the judgement of many; and I too for a long time was of the same opinion. The whole matter seemed to rest on sophistry rather than on truth. Yet I sought so long, until I actually proved this to be the case." Here's how he did it, beginning with the observation that if the Cardan formula solution is actually real then it must be that $\sqrt[3]{2 + \sqrt{-121}}$ and $\sqrt[3]{2 - \sqrt{-121}}$ are complex conjugates,[5] i.e., if a and b are some yet to be determined real numbers, where

$$\sqrt[3]{2 + \sqrt{-121}} = a + b\sqrt{-1},$$

$$\sqrt[3]{2 - \sqrt{-121}} = a - b\sqrt{-1},$$

then we have $x = 2a$, which is indeed real. The first of these two statements says that

$$2 + \sqrt{-121} = (a + b\sqrt{-1})^3.$$

From the identity $(m + n)^3 = m^3 + n^3 + 3mn(m + n)$, with $m = a$ and $n = b\sqrt{-1}$, we get

BOX 1.2

THE IRREDUCIBLE CASE MEANS THERE ARE THREE REAL ROOTS

To study the nature of the roots to $x^3 = px + q$, where p and q are both non-negative, consider the function

$$f(x) = x^3 - px - q.$$

Calculating $f'(x) = 3x^2 - p$, we see that the plot of $f(x)$ will have tangents with zero slope at $x = \pm\sqrt{p/3}$, i.e., the local extrema of the depressed cubic that can lead to the irreducible case are symmetrically located about the vertical axis. The values of $f(x)$ at these two local extrema are, if we denote them by M_1 and M_2,

$$M_1 = \frac{p}{3}\sqrt{\frac{p}{3}} - p\sqrt{\frac{p}{3}} - q = -\frac{2}{3}p\sqrt{\frac{p}{3}} - q, \text{ at } x = +\sqrt{\frac{p}{3}},$$

$$M_2 = -\frac{p}{3}\sqrt{\frac{p}{3}} + p\sqrt{\frac{p}{3}} - q = \frac{2}{3}p\sqrt{\frac{p}{3}} - q, \text{ at } x = -\sqrt{\frac{p}{3}}.$$

Notice that the local minima $M_1 < 0$, always (as p and q are both non-negative), while the local maximum M_2 can be of either sign, depending on the values of p and q. Now, if we are to have three real roots, then $f(x)$ must cross the x-axis three times and this will happen only if $M_2 > 0$, as shown in figure 1.2. That is, the condition for all real roots is $\frac{2}{3}p\sqrt{p/3} - q > 0$, or $\frac{4}{27}p^3 : q^2$, or, at last, $q^2/4 - p^3/27 < 0$. But this is precisely the condition ⌐ the Cardan formula that leads to imaginary numbers in the solution. That is, the occurrence of the irreducible case is always associated with three real roots to the cubic $f(x) = 0$. As the figure also makes clear, these three roots are such that two are negative and one is positive. See if you can also show that the sum of the three roots must be zero.*

* This is a particular example of the following more general statement. Suppose we write the nth-degree polynomial equation $x^n + a_{n-1}x^{n-1} + a_{n-2}x^{n-2} + \cdots + a_1x + a_0 = 0$ in factored form. That is, if we denote the n roots of the equation by $r_1, r_2, \ldots, r_n$, then we can write $(x - r_1)(x - r_2) \cdots (x - r_n) = 0$. By successively multiplying the factors together, starting at the left, you can easily show that the coefficient of the x^{n-1} term is the negative of the sum of the roots, i.e., that $a_{n-1} = -(r_1 + r_2 + \cdots + r_n)$. In the case of the depressed cubic, with no x^2 term, we have $a_2 = 0$ by definition, i.e., the sum of the roots of any depressed cubic equation is zero.

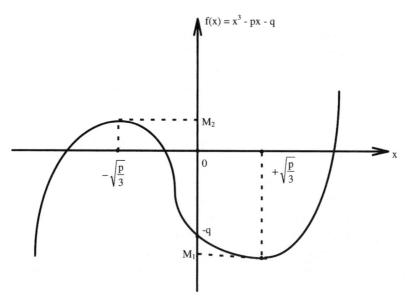

Figure 1.2. Plot of $f(x) = x^3 - px - q$, p and $q \geq 0$.

$$(a + b\sqrt{-1})^3 = a^3 - b^3\sqrt{-1} + 3ab\sqrt{-1}\,(a + b\sqrt{-1})$$
$$= a^3 - b^3\sqrt{-1} + 3a^2b\sqrt{-1} - 3ab^2$$
$$= a(a^2 - 3b^2) + b(3a^2 - b^2)\sqrt{-1}.$$

If this complex expression is to equal the complex number $2 + \sqrt{-121}$, then the real and imaginary parts must be separately equal, and so we arrive at the following pair of conditions:

$$a(a^2 - 3b^2) = 2,$$
$$b(3a^2 - b^2) = 11.$$

If we assume a and b are both integers (there is no a priori justification for this, but we are always free to try something and see where it goes), then perhaps you will notice that $a = 2$ and $b = 1$ work in both conditions. There are also ways to work this conclusion out more formally. For example, notice that 2 and 11 are prime, ask yourself what are the integer factors of any prime, and notice that if a and b are integers then so are $a^2 - 3b^2$ and $3a^2 - b^2$. For our purposes here, however, it is sufficient to see that

$$\sqrt[3]{2 + \sqrt{-121}} = 2 + \sqrt{-1},$$

$$\sqrt[3]{2 - \sqrt{-121}} = 2 - \sqrt{-1},$$

21

statements that are easily verified by cubing both sides. With these results Bombelli thus showed that the mysterious Cardan solution *is* $x = 4$, and this *is* correct. As shown in box 1.2, for the irreducible case with all three roots real, there is just one positive root; that is, the root given by the Cardan formula (see if you can prove this—read the last half of appendix A if you need help).

1.6 CALCULATING THE REAL ROOTS WITHOUT IMAGINARIES

Still, while the Cardan formula works in all cases, including the irreducible case, you might be wondering why there is not a formula that directly produces a real answer for the positive real root in the irreducible case. And, in fact, there is. Discovered by the great French mathematician Francoise Viète[6] (1540–1603), it gives all the roots of the irreducible cubic in terms of the cosine and arccosine (or inverse cosine) trigonometric functions. This discovery is all the more remarkable when one considers that Viète was not a professional mathematician, but rather was a lawyer in service to the state, under the kings Henri III and Henri IV. He did his mathematics when he could steal time away from his "more important" duties, such as decoding intercepted, encrypted letters written by the Spanish court during France's war with Spain. While clever, Viète's solution (published posthumously in 1615) seems not to be very well known, and so here is what he did.

Viète started his analysis with the cubic equation $x^3 = px + q$, with p and q written as $p = 3a^2$ and $q = a^2b$. That is, he started with the cubic

$$x^3 = 3a^2x + a^2b, \text{ with } a = \sqrt{\frac{p}{3}} \text{ and } b = \frac{3q}{p}.$$

Then, he used the trigonometric identity

$$\cos^3(\theta) = \frac{3}{4}\cos(\theta) + \frac{1}{4}\cos(3\theta).$$

If you don't recall this identity, just accept it for now—I will derive it for you in chapter 3 using complex numbers. Viète's next step was to suppose that one can always find a θ such that $x = 2a\cos(\theta)$. I'll now show you that this supposition is in fact true by actually calculating the required value of θ. From the supposition we have $\cos(\theta) = x/2a$, and if this is substituted into the above trigonometric identity then you can quickly show that $x^3 = 3a^2x + 2a^3\cos(3\theta)$. But this is just the cubic we are trying to solve if we write $2a^3\cos(3\theta) = a^2b$. That is,

$$\theta = \frac{1}{3}\cos^{-1}\left(\frac{b}{2a}\right).$$

Inserting this result for θ into $x = 2a\cos(\theta)$ immediately gives us the solution

$$x = 2a\cos\left\{\frac{1}{3}\cos^{-1}\left(\frac{b}{2a}\right)\right\}$$

or, in terms of p and q,

$$x = 2\sqrt{\frac{p}{3}}\cos\left\{\frac{1}{3}\cos^{-1}\left(\frac{3\sqrt{3}q}{2p\sqrt{p}}\right)\right\}.$$

For this x to be real, the argument of the $\cos^{-1}$ must be no greater than one, i.e., $3\sqrt{3}q \le 2p^{3/2}$. (Later in this book, in chapter 6, I will discuss what happens when the magnitude of the argument in the inverse cosine function *is* greater than one.) But this condition is easily shown to be equivalent to $q^2/4 - p^3/27 \le 0$, which is precisely the condition that defines the irreducible case. Notice that imaginary quantities do not appear in Viète's formula, unlike the Cardan formula.

Does Viète's formula work? As a test, recall Bombelli's cubic $x^3 = 15x + 4$, with $p = 15$ and $q = 4$. Viète's formula gives

$$x = 2\sqrt{5}\cos\left\{\frac{1}{3}\cos^{-1}\left(\frac{12\sqrt{3}}{30\sqrt{15}}\right)\right\}.$$

This rather fearsome-looking expression is easily run through a hand calculator to give $x = 4$, which is correct. This root is found by taking $\cos^{-1}(12\sqrt{3}/30\sqrt{15}) = 79.695°$. But a quick sketch of the cosine function will show that the angles $280.305°$ and $439.695°$ are equally valid. Evaluating x for these two angles will give the other real roots -0.268 and -3.732, i.e., $-2 \pm \sqrt{3}$. Viète himself, however, paid no attention to negative roots. And for another quick check, consider the special case when $q = 0$. Then, $x^3 - px = 0$ which by inspection has the three real roots $x = 0$, $x = \pm\sqrt{p}$. That is, $x = \sqrt{p}$ is the one positive root. Viète's formula gives, for $q = 0$,

$$x = 2\sqrt{\frac{p}{3}}\cos\left\{\frac{1}{3}\cos^{-1}(0)\right\} = 2\sqrt{\frac{p}{3}}\cos(30°)$$

since $\cos^{-1}(0) = 90°$. But $(2/\sqrt{3})\cos(30°) = 1$ and so Viète's formula does give $x = \sqrt{p}$. And since $\cos^{-1}(0) = 270°$ (and $450°$), too, you can easily verify that the formula gives the $x = 0$ and $x = -\sqrt{p}$ roots, as well. Technically, this is not an irreducible cubic, but Viète's formula still works. Notice that the roots in these two specific cases satisfy the last statement made in box 1.2.

Viète knew very well the level at which his analytical skills operated. As he himself wrote of his mathematics, it was "not alchemist's gold, soon to go up in smoke, but the true metal, dug out from the mines where dragons are standing watch." Viète was not a man with any false modesty. If his solution had been found a century earlier, would Cardan have worried much over the imaginaries that appeared in his formula? Would Bombelli have been motivated to find the "realness" of the complex expressions that appear in the formal solution to the irreducible cubic? It is interesting to speculate about how the history of mathematics might have been different if some genius had beaten Viète to his discovery. But there was no such genius, and it was Bombelli's glory to unlock the final secret of the cubic.

Bombelli's insight into the nature of the Cardan formula in the irreducible case broke the mental logjam concerning $\sqrt{-1}$. With his work, it became clear that manipulating $\sqrt{-1}$ using the ordinary rules of arithmetic leads to perfectly correct results. Much of the mystery, the near-mystical aura, of $\sqrt{-1}$ was cleared away with Bombelli's analyses. There did remain one last intellectual hurdle to leap, however, that of determining the *physical* meaning of $\sqrt{-1}$ (and that will be the topic of the next two chapters), but Bombelli's work had unlocked what had seemed to be an unpassable barrier.

1.7 A Curious Rediscovery

There is one last curious episode concerning the Cardan formula that I want to tell you about. About one hundred years after Bombelli explained how the Cardan formula works in all cases, including the irreducible case where all roots are real, the young Gottfried Leibniz (1646–1716) somehow became convinced the issue was still open. This is all the more remarkable because Leibniz is known to have studied Bombelli's *Algebra,* and yet he thought there was still something left to add to the Cardan formula. Leibniz was a genius, but this occurred at about age twenty-five, when, as one historian put it, "Leibniz had but little of any competence in what was then modern mathematics. Such firsthand knowledge as he had was mostly Greek."[7]

Leibniz had, at that time, just met the great Dutch physicist and mathematician Christian Huygens (1629–95), with whom he began a lifelong correspondence. In a letter written sometime between 1673 and 1675 to Huygens,[8] he began to rehash what Bombelli had done so long before. In this letter he communicated his famous (if anticlimactic) result

$$\sqrt{1+\sqrt{-3}} + \sqrt{1-\sqrt{-3}} = \sqrt{6},$$

of which Leibniz later declared, "I do not remember to have noted a more singular and paradoxical fact in all analysis; for I think I am the first one to have reduced irrational roots, imaginary in form, to real values. . . ." Of course, it was Bombelli who was the first, by a century.

When the imaginary number $\sqrt{-1}$ is first introduced to high school students it is common to read something like the following (which, actually, I've taken from a college level textbook[9]): "The real equation $x^2 + 1 = 0$ led to the invention of i (and also $-i$) in the first place. That was declared to be the solution and the case was closed." Well, of course, this is simple to read and easy to remember but, as you now know, it is also not true. When the early mathematicians ran into $x^2 + 1 = 0$ and other such quadratics they simply shut their eyes and called them "impossible." They certainly did not invent a solution for them. The breakthrough for $\sqrt{-1}$ came not from quadratic equations, but rather from cubics which clearly had real solutions but for which the Cardan formula produced formal answers with imaginary components. The basis for the breakthrough was in a clearer-than-before understanding of the idea of the conjugate of a complex number. Before continuing with Leibniz, then, let me show you a pretty use of the complex conjugate.

Consider the following statement, easily shown to be correct with a little arithmetic on the back of an envelope:

$$(2^2 + 3^2)(4^2 + 5^2) = 533 = 7^2 + 22^2 = 23^2 + 2^2.$$

And this one, which is only just a bit more trouble to verify:

$$(17^2 + 19^2)(13^2 + 15^2) = 256{,}100 = 64^2 + 502^2 = 8^2 + 506^2.$$

What is going on here?

These are two examples of a general theorem that says the product of two sums of two squares of integers is always expressible, in two different ways, as the sum of two squares of integers. That is, given integers a, b, c, and d, we can always find two pairs of positive integers u and v such that

$$(a^2 + b^2)(c^2 + d^2) = u^2 + v^2.$$

Therefore, says this theorem, it must be true that there are two integer solutions to, for example,

$$(89^2 + 101^2)(111^2 + 133^2) = 543{,}841{,}220 = u^2 + v^2.$$

Can you see what u and v are? Probably not. With complex numbers, and the concept of the complex conjugate, however, it is easy to analyze this problem. Here is how to do it.

Factoring the above general statement of the theorem to be proved, we have

$$[(a + ib)(a - ib)][(c + id)(c - id)] = (a + ib)(c + id)][(a - ib)(c - id)].$$

Since the quantities in the right-hand brackets are conjugates, we can write the right-hand side as $(u + iv)(u - iv)$. That is,

$$u + iv = (a + ib)(c + id) = (ac - bd) + i(bc + ad)$$

and so

$$u = |ac - bd| \text{ and } v = bc + ad.$$

But this isn't the only possible solution. We can also write the factored expression as

$$[(a + ib)(c - id)][(a - ib)(c + id)] = [u + iv][u - iv]$$

and so a second solution is

$$u + iv = (a + ib)(c - id) = (ac + bd) + i(bc - ad)$$

or,

$$u = ac + bd \text{ and } v = |bc - ad|.$$

These results prove the theorem by actually constructing formulas for u and v, and in particular they tell us that

$$(89^2 + 101^2)(111^2 + 133^2) = 3,554^2 + 23,048^2 = 626^2 + 23,312^2.$$

This problem is quite old (it was known to Diophantus), and a discussion of it, one *not* using complex numbers, can be found in the 1225 book *Liber quadratorum* (*The Book of Squares*[10]) by the medieval Italian mathematician Leonardo Pisano (circa 1170–1250), i.e., Leonardo of Pisa, a town best known today for its famous leaning tower. Leibniz, no doubt, would have found the concept of the complex conjugate to be just what was needed to explain his "paradoxical fact."

As Leibniz expressed his confusion, "I did not understand how . . . a quantity could be real, when imaginary or impossible numbers were used to express it." He found this so astonishing that after his death, among some unpublished papers, several such expressions were found, as if he had calculated them endlessly. For example, solving the cubics $x^3 - 13x - 12 = 0$ and $x^3 - 48x - 72 = 0$, respectively, led him to the additional discoveries that

$$\sqrt[3]{6 + \sqrt{-\frac{1225}{27}}} + \sqrt[3]{6 - \sqrt{-\frac{1225}{27}}} = 4$$

and

$$\sqrt[3]{36 + \sqrt{-2800}} + \sqrt[3]{36 - \sqrt{-2800}} = -6.$$

The realness of the literally complex expressions on the left would, today, be considered by a good high school algebra student to be trivially obvious. Such has been the progress in mathematics in understanding $\sqrt{-1}$. Indeed, using the conjugate concept, we know today that a plot of *any* function $f(x)$ contains, in its very geometry, *all* the roots to the equation $f(x) = 0$, real *and* complex. Let me conclude this chapter by showing you how this is so, in particular, for quadratics and for cubics.

1.8 How to Find Complex Roots with a Ruler

When an nth-degree polynomial $y = f(x)$ with real coefficients is plotted, the geometrical interpretation is that the plot will cross the x-axis once for each real root of the equation $f(x) = 0$. Crossing the x-axis is, in fact, where the zero on the right-hand side comes from. If there are fewer than n crossings, say $m < n$, then the interpretation is that there are m real roots, given by the crossings, and $n - m$ complex roots. The value of $n - m$ is an even number since, as shown in appendix A, complex roots always appear as conjugate *pairs*. This is not to say, however, that there is no physical signature in the plot for the complex roots. The signature for the real roots, the x-axis crossings, is simple and direct, but if you are willing to do just a bit more work you can read off the complex roots, too.

First, consider the quadratic $f(x) = ax^2 + bx + c = 0$. The two roots to this equation are either both real or a complex conjugate pair, depending on the algebraic sign of the quantity $b^2 - 4ac$. If this is non-negative, then the roots are real and there are either two x-axis crossings or a touching of the axis (if $b^2 - 4ac = 0$, giving a double root). If $b^2 - 4ac$ is negative then the roots are complex and there are no x-axis crossings, which is the case shown in figure 1.3. Let us assume that this is the case, and that the roots are $p \pm iq$. Then, writing $f(x)$ in factored form,

$$f(x) = a(x - p - iq)(x - p + iq) = a[(x - p)^2 + q^2],$$

it is clear that $f(x) \geq aq^2$ if $a > 0$, and $f(x) \leq aq^2$ if $a < 0$. That is, $f(x)$ takes on its minimum value at $x = p$ if $a > 0$ (as shown in figure 1.3), or its

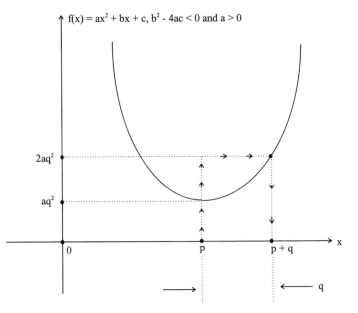

Figure 1.3. A quadratic equation with no real roots.

maximum value at $x = p$ if $a < 0$. We can, therefore, measure p from the plot of $f(x)$ as the x-coordinate of the local extremum.

Next, to measure the value of q from the plot, first measure the y coordinate of the minimum (I'm assuming $a > 0$, but the $a < 0$ case is a trivial variation), i.e., measure aq^2. Then, at $x = p$, first move upward $2aq^2$, then over to the right until you intersect the plot. The x-value of this intersection point (call it $\hat{x}$), when plugged into the quadratic equation, gives

$$f(\hat{x}) = 2aq^2 = a[(\hat{x} - p)^2 + q^2] = a(\hat{x} - p)^2 + aq^2$$

or,

$$aq^2 = a(\hat{x} - p)^2 \text{ or } q = \hat{x} - p.$$

Thus, q can be directly measured off the plot of $f(x)$, as shown in figure 1.3.

Concentrating next on cubics, observe first that there will be either (a) three real roots or (b) one real root and two complex conjugate roots. Be sure you are clear in your mind why all three roots cannot be complex, and why there cannot be two real roots and one complex root. If you're not clear on this, see appendix A. Case (b) is the one of interest for us. Call the real root $x = k$, and the conjugate pair of roots $x = p \pm iq$. Then, we can write $f(x)$ in factored form as

$$y = f(x) = (x - k)(x - p + iq)(x - p - iq)$$

or, expanding and collecting terms, as

$$f(x) = (x - k)(x^2 - 2xp + p^2 + q^2).$$

The plot of a cubic with a single real root, which means one x-axis crossing, will have the general appearance of figure 1.4. Construct the triangle AMT, where A is the intersection point of $y = f(x)$ with the x-axis, T is the point of tangency to $y = f(x)$ of a straight line passing through A, and M is the foot of the perpendicular through T perpendicular to the x-axis. Of course, the real root is $k = OA$.

Now, consider the straight line $y = \lambda(x - k)$, which clearly passes through A as $y = 0$ when $x = k$. Imagine that the slope of this line, λ, is adjusted until it just touches $y = f(x)$, i.e., until it is tangent to $y = f(x)$. This then gives us T, and since the x-value at T is common to both $y = f(x)$ and $y = (x - k)$, calling this x-value $\hat{x}$ says that

$$\lambda(\hat{x} - k) = (\hat{x} - k)(\hat{x}^2 - 2p\hat{x} + p^2 + q^2).$$

Since $\hat{x} - k \neq 0$, we can divide through both sides of this equation to obtain a quadratic in $\hat{x}$,

$$\lambda = \hat{x}^2 - 2p\hat{x} + p^2 + q^2.$$

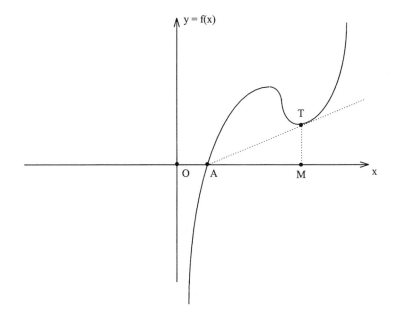

Figure 1.4. A cubic equation with one real root.

In fact, since T is a point of tangency, there must be exactly one value of $\hat{x}$. That is,

$$\hat{x}^2 - 2p\hat{x} + p^2 + q^2 - \lambda = 0$$

must have two equal, or double, roots. Now, in general,

$$\hat{x} = \frac{2p \pm \sqrt{4p^2 - 4(p^2 + q^2 - \lambda)}}{2},$$

and to have double roots the radical must be zero. That is,

$$4p^2 - 4(p^2 + q^2 - \lambda) = 0$$

or, $\lambda = q^2$. That is, the tangent line AT has slope $q^2 = TM/AM$. The value of $\hat{x}$ is, from the general expression for $\hat{x}$, then just $\hat{x} = p = OM$.

So, to find all the roots of the cubic, you need only plot $y = f(x)$ and then:

1. Read off the real root by measuring OA ($= k$).
2. Place a straightedge at A as a pivot point and swing the edge slowly until it just touches the plotted function (thus "locating" T).
3. Measure TM and AM, and then calculate

$$q = \sqrt{\frac{TM}{AM}}.$$

4. Measure OM to give p.
5. The two imaginary roots are $p + iq$ and $p - iq$.

A First Try at Understanding the Geometry of $\sqrt{-1}$

2.1 RENÉ DESCARTES

Despite the success of Bombelli in giving formal meaning to $\sqrt{-1}$ when it appeared in the answers given by Cardan's formula, there still lacked a physical interpretation. Mathematicians of the sixteenth century were very much tied to the Greek tradition of geometry, and they felt uncomfortable with concepts to which they could not give a geometric meaning. This is why, two centuries after Bombelli's *Algebra,* we find Euler writing in his *Algebra* of 1770

> All such expressions as $\sqrt{-1}$, $\sqrt{-2}$, etc., are consequently impossible or imaginary numbers, since they represent roots of negative quantities; and of such numbers we may truly assert that they are neither nothing, nor greater than nothing, nor less than nothing, which necessarily constitutes them imaginary or impossible.

No such "negative" feelings were directed toward the square roots of positive numbers, however, because, at least in part, it was known that a geometric construction of them could be given. The following construction is given by René Descartes (1596–1650) in his 1637 *La Geometrie.*[1] Suppose GH is a given line segment as shown in figure 2.1, and the problem is to construct another line segment with length equal to $\sqrt{GH}$. Descartes began by extending GH to F, where FG is of unit length. FG has to be a given length, too, and it establishes the size or *scale* of the construction. Thus, $FH = FG + GH = 1 + GH$. Next, he used the well-known method for bisecting a line segment to locate K, the midpoint of FH. Then, using K as the center, he constructed the semicircle FIH with radius $KH = FK$. Finally, at G he erected a perpendicular that intersects the semicircle at I (and so IK is a radius, too). From all this we can write that

$$FG + GH = 2IK,$$

$$1 + GH = 2IK,$$

$$\frac{1}{2}(1 + GH) = IK.$$

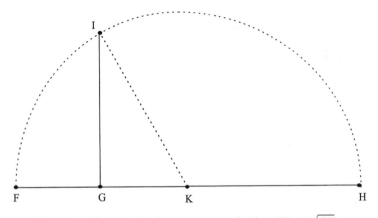

Figure 2.1. Constructing the square root of a line ($IG = \sqrt{GH}$).

Also,

$$FG + GK = IK,$$

$$GK = IK - FG = IK - 1 = \frac{1}{2}(1 + GH) - 1,$$

$$GK = \frac{1}{2}(GH - 1).$$

Now, from the Pythagorean theorem we have

$$(IG)^2 + (GK)^2 = (IK)^2,$$

$$(IG)^2 + \frac{1}{4}(GH - 1)^2 = \frac{1}{4}(1 + GH)^2,$$

$$(IG)^2 = \frac{1}{4}[(1 + GH)^2 - (GH - 1)^2] = GH$$

Thus, $IG = \sqrt{GH}$. Descartes himself wrote none of this, but he did write, as the very last line of *La Geometrie*, words that show his omissions were no oversight: "I hope that posterity will judge me kindly, not only as to things which I have explained, but also as to those which I have intentionally omitted so as to leave to others the pleasure of discovery." A man with a sense of irony.

It is important to distinguish between the construction of the square root of *any* given line segment, as described by Descartes, and the construction of a line segment equal to a specified square root. The second case, for the square root of integer lengths (assuming, once again, that one has a priori knowledge of the unit length), was given an elegant geometric solution long before Descartes by the fourth-century B.C. mathematician Theodorus of Cyrene. Theodorus, who taught Plato mathematics, demonstrated that the square roots of all the nonsquare integers from 3 to 17 are irrational—his student Theaetetus extended this result to all nonsquare integers. All of Theodorus' work is lost, and it is not known how he arrived at his results, but we know of them because Plato put both men into his dialogue *Theaetetus* and tells us of their individual achievements.

It has been suggested that perhaps Theodorus' solution was based on the following method for constructing $\sqrt{n}$ for any positive integer $n > 1$, as shown in figure 2.2. This figure shows a spiral of right triangles with a common vertex, where in each triangle the side opposite the common vertex has

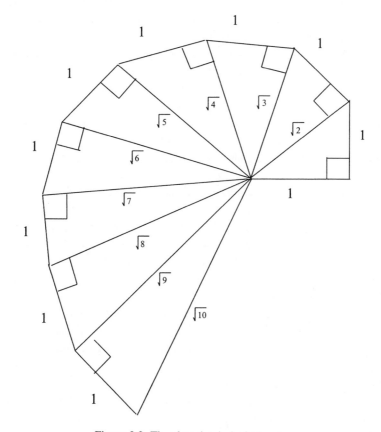

Figure 2.2. Theodorus' spiral of triangles.

length 1. The hypotenuse of the nth triangle then has length $\sqrt{n+1}$. Plato wondered why Theodorus stopped with $\sqrt{17}$ in his irrationality analyses ("somehow he got into difficulties"), but figure 2.2 may provide the reason. If we compute the sum of the vertex angles for the first n triangles, then we have

$$\tan^{-1}\left(\frac{1}{\sqrt{1}}\right) + \tan^{-1}\left(\frac{1}{\sqrt{2}}\right) + \tan^{-1}\left(\frac{1}{\sqrt{3}}\right) + \ldots + \tan^{-1}\left(\frac{1}{\sqrt{n}}\right).$$

For $n = 16$ (which gives $\sqrt{17}$) this sum is $351.15°$, while for $n = 17$ the sum is $364.78°$. That is, perhaps Theodorus stopped at $\sqrt{17}$ simply because for $n > 16$ his spiral started to overlap itself and the drawing became "messy."

These constructions deal with the square roots of *positive* lengths. But what, geometrically, could the square root of a negative mean? According to Descartes it meant the sheer impossibility of doing a geometric construction. And just who was this Descartes, you must by now be wondering? Born into the lower levels of French nobility, Descartes received a good general education from Jesuits, spent two years in Paris studying mathematics by himself, and wound up in 1617 as a twenty-one year old gentleman officer in military service to a prince. Two years later he quit the military life because, as he recounted later, he had had several dreams in which tantalizing ideas were revealed to him that led eventually to his work in analytic geometry. A possible loss to the brotherhood of warriors, maybe, but it was a decision of inestimable gain for mathematics. In 1628 Descartes moved to Holland where he lived a mostly solitary life as a scholar. In 1637 he used his knowledge of geometry to develop the first scientific explanation of the rainbow, based on tracing rays of light through one (or more) internal reflections of the rays inside water drops. In 1649 he moved to Stockholm to tutor Queen Christina. The next year the brutal Swedish winter did him in, and he died of pneumonia.

To see how Descartes understood the association of imaginary numbers with geometric impossibility, consider his demonstration on how to solve quadratic equations with geometric constructions (what follows is from his *La Geometrie*). He began with the equation $z^2 = az + b^2$, where a and b^2 are both non-negative, and which he took to be the lengths of two given line segments. In particular, suppose as shown in figure 2.3 that LM is equal to the square root of the given b^2. Suppose, too, that $LN = \frac{1}{2}a$, an easy bisecting construction of the given a, and that LN is erected perpendicular to LM. Next, construct the circle of radius $\frac{1}{2}a$, centered on N, draw the line NM, and, finally, extend

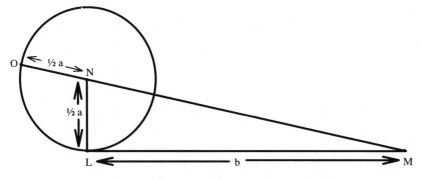

Figure 2.3. Descartes' geometric construction of the positive root to $z^2 = az + b^2$, with a and b^2 both positive.

NM to intersect the other side of the circle at O. Then it is immediately obvious that

$$OM = \frac{1}{2}a + \sqrt{\left(\frac{1}{2}a\right)^2 + b^2},$$

which is the positive algebraic solution to the quadratic $z^2 = az + b^2$. Thus, Descartes has geometrically constructed one solution to the quadratic. This construction always works, for any given positive values of a and b^2. Notice that Descartes is ignoring the other solution of $z = \frac{1}{2}a - \sqrt{\frac{1}{4}a^2 + b^2}$, which for any positive a and b^2 is always negative. His reason for doing this was, as I have stressed before, that the mathematicians of his day did not accept such *false* roots, as Descartes called them.

Descartes next considered the quadratic $z^2 = az - b^2$. The algebraic solution is

$$z = \frac{1}{2}a \pm \sqrt{\frac{1}{4}a^2 - b^2},$$

and so now *complex* roots are possible, even with a and b^2 constrained to be positive. Descartes explored the geometric implications of this possibility, as follows. He started, as before, with the same line segments $LN = \frac{1}{2}a$ and $LM = b$. But instead of joining N and M, he erected a perpendicular line at M and then, with N as the center (see figure 2.4), he drew the circle with radius $\frac{1}{2}a$. The two points where this circle intersects the perpendicular line, if such intersections occur, define the two points Q and R. Descartes then observed that the

35

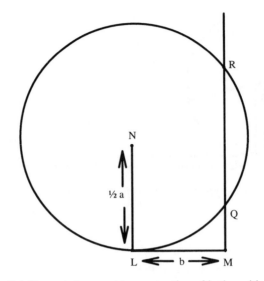

Figure 2.4. Descartes' geometric construction of both positive roots to $z^2 = az - b^2$, with a and b^2 both positive.

two line segments MQ and MR are the two solutions to the quadratic, if they exist. This "observation" is a pretty problem in algebra and geometry, and you should try your hand at showing that MQ and MR are, indeed, equal to the two values of z given above—but *try* before you look at the hint.[2] Notice that now Descartes is happy to acknowledge *both* solutions given by his geometric construction, since if the two values $z = \frac{1}{2}a \pm \sqrt{\frac{1}{4}a^2 - b^2}$ are real then they are also positive.

What did Descartes conclude from all this? As he wrote at the end of his analysis, "And if the circle described about N and passing through L neither cuts nor touches the line MQR, the equation has no root [*real* root, is what Descartes should have said], and so *we may say that the construction of the problem is impossible* [my emphasis]." Descartes thus excludes the case of a double root when he does not allow the circle just to touch the perpendicular, i.e., to let R and Q be one and the same point. Notice that, for there to be no intersection (or "touching") at all, the geometric condition is $b > \frac{1}{2}a$, which is precisely the algebraic condition that gives complex roots to the quadratic.

I am ignoring Descartes' treatment of the case of $z^2 + az = b^2$ because it adds nothing to the discussion here. That quadratic, and the two I discussed above, are the only ones presented by Descartes because there is always at least one positive root to them. He completely ignores the fourth possibility of

$z^2 + az + b^2 = 0$ because it never has positive roots if a and b^2 are both positive, the only sort of root that, for Descartes, had geometric significance. Associating the appearance of imaginary numbers with the physically impossible is a routine concept to a modern engineer or physicist, however, and it is not difficult to construct a simple physics example of how this can happen.

Imagine that a man is running at his top speed of v feet per second, to catch a bus that is stopped at a traffic light. When he is still a distance of d feet from the bus, the light changes and the bus starts to move away from the running man with a constant acceleration of a feet per second per second. When will the man catch the bus? To answer this question, let $x = 0$ represent the position of the light (the light will serve as the origin of our coordinate system), let x_b denote the position of the bus, and let x_m denote the position of the man. Then, at time $t = 0$, for example, $x_b = 0$ and $x_m = -d$, i.e., $t = 0$ is the instant the light changes. For an arbitrary time $t \geq 0$ we can write

$$x_b = \frac{1}{2}at^2,$$
$$x_m = -d + vt.$$

If we imagine that the man catches the bus at time $t = T$, then from the meaning of *catching* we have $x_b(T) = x_m(T)$, i.e.,

$$\frac{1}{2}aT^2 = -d + vT,$$

a quadratic equation for T. The solution is

$$T = \frac{v}{a} \pm \sqrt{\left(\frac{v}{a}\right)^2 - 2\frac{d}{a}} = \frac{v}{a} \pm i\sqrt{2\frac{d}{a} - \left(\frac{v}{a}\right)^2}.$$

If $d > \frac{1}{2}v^2/a$ then T is a complex time, and this is interpreted to mean that the man cannot actually catch the bus. But that does not mean there is no physically significant meaning to T. To see this, write $s = x_b - x_m$, i.e., s is the separation between the bus and the man. Thus,

$$s = \frac{1}{2}at^2 + d - vt.$$

To *catch* the bus at time T means that then $s = 0$. We can now ask a new question related to the question of catching the bus. We suppose that the man does not catch the bus, but now we ask at what time is the man *closest* to the bus? In other words, at what time is s minimum? Setting $ds/dt = 0$ gives $t = v/a$.

That is, the man is closest to catching the bus at a time equal to the real part of the complex time T. The imaginary part of T has physical significance, too, albeit a different one. The physical distinction between catching and not catching the bus is equivalent to T being real or complex. This latter distinction is controlled by the imaginary part of T, i.e., the transition between catching and not catching the bus occurs at the condition $2d/a = (v/a)^2$. Given any two of the three quantities d, a, and v, we can use this condition to determine the critical value of the third quantity that determines whether or not the man catches the bus. For example, if we are given both v and a, then d must be no greater than $\frac{1}{2}v^2/a$ if the man is to catch the bus.

Finally, here's one more question for you to think about. Suppose the man does catch the bus, which means T is real. For which T does this occur—there are, after all, *two* positive values given for T, determined by the $\pm$ signs? That is, what is the physical significance of two real roots? See the answer[3] only after thinking about this for a while.

It is not too hard to construct a more mathematical example. Imagine, as shown in figure 2.5, the circle with radius one centered on the origin described by the equation $x^2 + y^2 = 1$. Consider the point $(0,b)$ on the y-axis, where $b >$

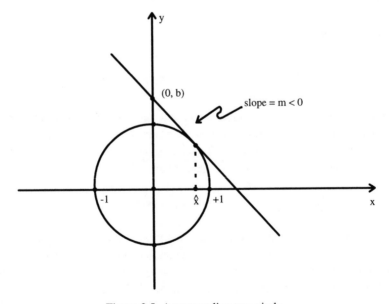

Figure 2.5. A tangent line to a circle.

1, which means the point is outside the circle, as shown. Suppose we draw the line through that point that is also tangent to the circle and which intersects the positive x-axis, as shown. What is the slope of that tangent line? If we call the slope m then the equation for the tangent line is that old workhorse from high school geometry, $y = mx + b$. By inspection, we have $m < 0$. At the point of tangency the line and the circle share that point, and so *at that point* $x^2 + (mx + b)^2 = 1$. That is, if $\hat{x}$ is the x-coordinate of the point of tangency, then $x = \hat{x}$ is the solution to this quadratic. We can expand the quadratic out and write its solution, using the quadratic formula:

$$\hat{x} = \frac{-2mb \pm \sqrt{4m^2b^2 - 4(m^2 + 1)(b^2 - 1)}}{2(m^2 + 1)}.$$

Since there must be exactly *one* solution—by definition, a tangent line just touches the circle and does not cut through it at multiple points—then the expression under the square root sign must equal zero. Solving for m and remembering it must be negative, we have

$$m = -\sqrt{b^2 - 1}$$

and

$$\hat{x} = \sqrt{1 - \left(\frac{1}{b}\right)^2}.$$

These expressions are real for $b > 1$, as we supposed, but what if $0 < b < 1$, i.e., what if the point on the y-axis is inside the circle? It is then physically obvious that we can no longer actually draw a tangent line to the circle, and that the expressions for m and $\hat{x}$ now give imaginary values. Here we see a direct connection between the onset of imaginary numbers and our physical inability to make a geometric construction.[4]

Finally, now that the idea of a line with an imaginary slope has come up, let me point out a curious property that such lines can have. Suppose we have two straight lines making angles α and β with the x-axis, as shown in figure 2.6. The tangents of these angles are, of course, the slopes of the lines. That is, if $m = \tan(\alpha)$ and $n = \tan(\beta)$, then the equations of the two lines are $y = nx + b_1$ and $y = mx + b_2$. Now, the angle between the two lines is $\phi = \beta - \alpha$ and, using a result from trigonometry,

$$\tan(\phi) = \tan(\beta - \alpha) = \frac{\tan(\beta) - \tan(\alpha)}{1 + \tan(\beta)\tan(\alpha)} = \frac{n - m}{1 + nm}.$$

For example, suppose the two lines are parallel; then $m = n$ and $\tan(\phi) = 0$, which says $\phi = 0$, i.e., parallel lines are inclined at angle zero to each other.

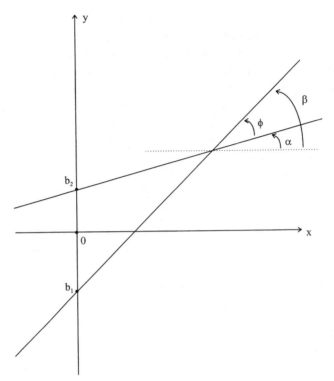

Figure 2.6. Two straight lines that cross in a plane.

This is all quite obvious, right? Well, suppose that we have two lines, each with the same *imaginary* slope i. Then,

$$\tan(\phi) = \frac{i-i}{1+i^2} = \frac{0}{1-1} = \frac{0}{0},$$

which is *indeterminate*. Notice that this weird result occurs only for an imaginary slope of i. If $m = n = ki$ with $k \neq i$, then $\tan(\phi) = 0$ as it "should." What is so special about a slope of precisely i? I don't know—perhaps it is best, for now, just to enjoy the mystery of this calculation for its own sake.

2.2 JOHN WALLIS

Despite Descartes' association of *imaginary* with the *impossibility of a geometric construction,* at least one younger mathematician thought that there might be something to construct that would represent $\sqrt{-1}$. John Wallis, mentioned in the previous chapter, was a child prodigy who at age fourteen (in 1630, seven years before *La Geometrie* was published) both read and spoke Latin and read Greek, Hebrew, and French. Ironically, he only began to study arithmetic as a "pleasing diversion in spare hours" the following year. Wallis made rapid progress after that, however, and by 1647 or 1648 he had progressed sufficiently far that he was able to rediscover the Cardan formula for himself. Still, his underlying training was as a theologian and he served as royal chaplain to Charles II, and declined an offer in 1692 from Queen Mary II to make him a deacon.

It therefore was something of a surprise when, in 1649, he was appointed Savilian Professor of Geometry at Oxford University. This appointment almost surely came as a reward for his valuable services to Parliament as a decipherer of captured coded messages during its struggle with King Charles I and his Royalist followers. Wallis was not a sycophant, however, because while he was favored by Parliament he did not curry such favor—he was one of the signers of the petition to Parliament not to execute the deposed Charles I, and he signed *before* Cromwell had Wallis appointed to Oxford.

Wallis was one of the original members of a group of like-minded men at Oxford that evolved into the Royal Society of London, and he eventually served as its President. His 1655 book *Arithmetica Infinitorum* contains the germ of the integral calculus; for example, it discusses the area under curves of the form $y = x^n$, and it contains Wallis' famous product formula for π,

$$\frac{\pi}{2} = \frac{2}{1} \cdot \frac{2}{3} \cdot \frac{4}{3} \cdot \frac{4}{5} \cdot \frac{6}{5} \cdot \frac{6}{7} \cdot \frac{8}{7} \cdot \frac{8}{9} \cdots ,$$

which alternately over- and underestimates the value of π as more and more factors are included. In chapter 6 I will show you how to derive this. Wallis' book is known to have had great influence on the young Isaac Newton.

The connection between Wallis and Newton was renewed many years later when, in the controversy between Newton and Leibniz over which man had priority in developing the differential calculus, Wallis as President of the Royal Society became involved in mediating the conflicting claims of the two. After his death he was interred in the Oxford University church, and the following words were inscribed on a nearby wall: "Here sleeps John Wallis, Doctor of Theology, Savilian Professor of Geometry, and Keeper of the Oxford Archives. He left immortal works. . . ." It is not surprising, then, that when a man of this caliber turned his attention to $\sqrt{-1}$ something of interest was the result.

There is evidence that years before his 1685 *Algebra,* in which he formally presented his analyses of imaginaries, Wallis had pondered and puzzled over the meaning of imaginary numbers in geometry. For example, Wallis engaged in a correspondence with the English mathematician John Collins (1625–83) on this topic, and typical of the problems he was considering is the one Collins related (in a letter dated October 19, 1675) to the Scottish mathematician James Gregory (1638–75). Collins wrote that in an earlier letter, to show how a "tiro" might be misled,[5] Wallis had analyzed a triangle with sides 1 and 2, on a base of 4. Wallis showed that if one simply goes through the formal algebra then the two segments of the base beneath the sides of length 1 and 2 come out real, even though the triangle is an impossible one (try drawing it!). Apparently Wallis did not pursue this "paradox" at that time because, in his letter to Gregory, Collins went on to write "but if he had proceeded further he would have found the perpendicular to have been [imaginary], which had manifested the impossibility." This seems to indicate that Wallis, before 1675, was perhaps still unsure of the precise connection between algebraic imaginaries and geometry. By 1685, however, Wallis had made progress.

As a prelude to his analyses of the square root of a negative number, Wallis began by observing, in his *Algebra,* that just negative numbers themselves, so long viewed with suspicion by mathematicians, did in fact have a perfectly clear physical interpretation. Directing his readers' attention to a straight line, with some point marked as the *zero point* or origin, Wallis wrote that a positive number means distance measured from the zero point *to the right,* and that a negative number means distance measured from the zero point *to the left.* In Wallis' own words, "And though, as to the bare Algebraick Notation, it [a negative number] import a quantity less than nothing: Yet, when it comes to a Physical Application, it denotes as Real a Quantity as if the Sign were +; but to be interpreted in a contrary sense." This all surely is obvious to a modern reader, but everything, even the obvious, is earned with hard thinking by some pioneering genius.

With the proper physical interpretation of negative numbers thus settled, Wallis then moved on to the real, or perhaps I should say imaginary, game he was after. He reminded his readers of the so-called *mean proportional,* i.e., if b and c are two given positive numbers, then x is the mean proportional of b and c if it satisfies the statement "b is to x as x is to c." Or, in algebra,

$$\frac{b}{x} = \frac{x}{c}$$

which says $x = \sqrt{bc}$. Sometimes b, x, and c are said to be in *harmonic progression* because three equally taut strings of length b, x, and c will vibrate and emit audible tones at frequencies that differ by equal intervals. Today you are much more likely to encounter the term *geometric mean* for *x;* the arithmetic mean is, of course, $(b + c)/2$. In the spirit of Descartes' *La Geometrie,* which Wallis had studied and greatly admired, Wallis showed how to geometrically construct the mean proportional of two given line segments. Here is how he did that.

As shown in figure 2.7, Wallis placed his zero point at *A* and then drew the line segment *AC.* Then, finding the midpoint of *AC,* he drew the circle with *AC* as a diameter. Next, picking the arbitrary point *B* on the diameter anywhere between *A* and *C,* Wallis erected the perpendicular at *B* that intersects the circle at *P.* It is clear that the two triangles *ABP* and *PBC* are right triangles, and so we can write

$$(BP)^2 + (AB)^2 = (AP)^2,$$
$$(BP)^2 + (BC)^2 = (PC)^2.$$

It is also true, but perhaps not so obvious, that *APC* is a right triangle, too.[6] So,

$$(AP)^2 + (PC)^2 = (AC)^2.$$

Substituting the first expressions for $(AP)^2$ and $(PC)^2$ into the third expression, and writing $AC = AB + BC$, then

$$(BP)^2 + (AB)^2 + (BP)^2 + (BC)^2 = (AB + BC)^2.$$

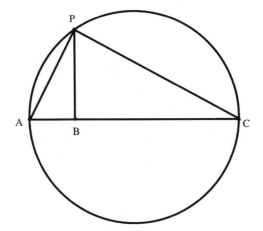

Figure 2.7. Wallis' construction of the mean proportional $[BP = \sqrt{(AB)(BC)}]$.

Expanding the right-hand side and collecting terms, this last expression reduces to $(BP)^2 = (AB)(BC)$, i.e., to

$$BP = \sqrt{(AB)(BC)},$$

which says BP is the mean proportional of AB and BC.

Wallis next modified this geometric construction of the mean proportional
of two positive line segments—positive because AB is extended to the right
from A to B, and BC is extended to the right from B to C—to include the case
of one of the line segments representing a negative value. To do this he drew
figure 2.7 again, up to the step of locating B and the perpendicular through B
intersecting the circle at P as shown in figure 2.8. But then, using P as a point
of tangency to the circle, Wallis extended the tangent line—that is, he constructed the line perpendicular to the radius PR (point R is the center of the
diameter AC) that passes through P— until it intersected the extension of the
diameter AC at the point B' to the left of A.

Now, by construction the triangle PRB' is a right triangle, and so from
Pythagoras we have

$$(B'P)^2 + (RP)^2 = (B'R)^2.$$

Since RP is a radius, then $RP = \frac{1}{2}AC$. Noticing, too, that $B'C = B'A + AC$, we
can then write $AC = B'C - B'A$ and so

$$RP = \frac{1}{2}(B'C - B'A) = \frac{1}{2}AC.$$

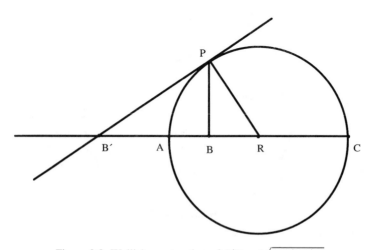

Figure 2.8. Wallis' construction of $B'P = \sqrt{(AB')(B'C)}$.

Also, $B'R = B'A + AR$, and, as AR is a radius, then $B'R = B'A + \frac{1}{2}AC$, or $B'R = B'A + \frac{1}{2}(B'C - B'A)$, or $B'R = \frac{1}{2}(B'A + B'C)$.

Now, substituting these expressions for RP and $B'R$ into the "Pythagoras" equation written for the PRB' triangle,

$$(B'P)^2 + \frac{1}{4}[(B'C) - (B'A)]^2 = \frac{1}{4}[(B'A) + (B'C)]^2$$

or, after expanding and collecting terms, $(B'P)^2 = (B'A)(B'C)$. If we suppose that the direction of extension of a line segment is irrelevant then we can just as well write this last result as $(B'P)^2 = (AB')(B'C)$, which gives

$$B'P = \sqrt{(AB')(B'C)},$$

a result that, except for the primes, looks just like the result derived for figure 2.5. But if we use Wallis' idea that direction *does* matter, then we have $B'C > 0$ and $AB' < 0$, and so $B'P$ is the square root of a *negative* quantity.

Well, what can one say about all this? It is clever, yes, but it certainly has the flavor of an exercise in splitting hairs at least as much as it has of geometry. Somehow it seems unlikely that we could really be saying very much, if anything, profound about $\sqrt{-1}$ with this sort of argument. As one twentieth-century author wrote of yet another attempt by Wallis at geometrically explaining $\sqrt{-1}$, it is "ingenious but scarcely convincing."[7] In fact, to his credit, Wallis himself wasn't terribly happy with this, either, and for him it was really just a warm-up. It was with yet another, completely different construction that he literally began to move in the correct direction to unlock the geometric secret of $\sqrt{-1}$.

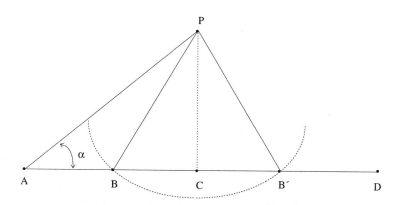

Figure 2.9. Two triangles that have two specified sides and an angle not included between the given sides.

Wallis began anew by treating a classic problem in geometry, that of constructing a triangle when two sides and an angle not included between the two sides, are given. This is often called an ambiguous problem because, as shown in figure 2.9, there are generally two possibilities. In the figure the two given sides are AP and PB ($=PB'$) and $\angle\, PAD = \alpha$ is the given angle. It is clear that the altitude (PC) of the triangle is determined and, as long as $PB = PB' > PC$, there are two solutions, the triangles APB and APB'. But if $PB = PB' < PC$ then there is no solution if we insist that solutions be upon the base AD (i.e., that B and B' be on AD).

Algebraically, what is going on can be expressed by the following two equations (remembering that $BC = CB'$):

$$AB = AC - BC = \sqrt{(AP)^2 - (PC)^2} - \sqrt{(PB)^2 - (PC)^2},$$
$$AB' = AC + CB' = \sqrt{(AP)^2 - (PC)^2} + \sqrt{(PB)^2 - (PC)^2}.$$

These two equations say that if $PC > PB$ then we get square roots of negative numbers, which Descartes would have interpreted as saying that the geometric construction of the desired triangle is impossible. But maybe that is not actually so.

Wallis' great insight was to realize that, even in the case of $PC > PB$, the given data still determine two points B and B' *if we allow them to be somewhere other than on the base AD.* Here's how he did it. Looking at figure 2.10, he constructed the circle with PC as a diameter. Then, using P as the center, Wallis swung another arc of radius PB until it intersected the first circle at B

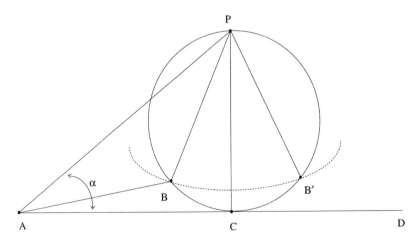

Figure 2.10. Wallis' construction hinting at *perpendicularity* as the meaning of an imaginary number.

and B'. He then argued that the triangles PAB (shown in the figure) and PAB' (not outlined) are the solution triangles, i.e., they are determined by the given sides AP and PB ($=PB'$) and by the angle $\angle PAD = \alpha$. The difference now, of course, is that the angle $\angle PAD$ does not appear in the solution triangle; but observe carefully that this was not a requirement in the original statement of the construction problem—only that the solution triangle be determined by two given sides and an angle. The big difference now is that the points B and B' do not lie on the base AD, but rather above it. Wallis had stumbled on the idea that, in some sense, the geometrical manifestation of imaginary numbers is *vertical* movement in the plane. Wallis himself made no such statement, however, and this is really a retrospective comment made with the benefit of three centuries of hindsight.

It would be another century before the now "obvious" representation of complex numbers as points in the plane, with the horizontal and vertical directions being the real and imaginary directions, respectively, would be put forth, but Wallis came very close. Close enough that at the beginning of the twentieth century we find the philosopher Ernst Mach—whose views are known to have greatly influenced Einstein—writing in his 1906 book *Space and Geometry* of $\sqrt{-1}$ as "a mean direction · proportional between $+1$ and -1." Even so, a close miss is still a miss, and Wallis' work in complex number geometry is remembered today only by historians.

The Puzzles Start to Clear

3.1 CASPAR WESSEL SEES THE WAY

More than a hundred years after Wallis' valiant but flawed attempt to tame complex numbers geometrically, the problem was suddenly and quite undramatically solved by the Norwegian[1] Caspar Wessel (1745–1818). This is both remarkable and, ironically, understandable, when you consider that Wessel was not a professional mathematician but a surveyor. Wessel's breakthrough on a problem that had stumped a lot of brilliant minds was, in fact, motivated by the practical problems he faced every day in making maps, i.e., by the survey data he regularly encountered in the form of plane and spherical polygons. There was no family tradition in mathematics to guide him, as both his father and his father's father were men of the cloth. It was Wessel's *work* that inspired him to success where all others had failed before.

Even though Wessel was one of thirteen children, and consequently the family finances must have been limited, he did receive a good high school education, followed by a year at the University of Copenhagen. He then left academic life and began his work as a cartographer while still a teenager (1764), as an assistant with the Danish Survey Commission operated by the Royal Danish Academy of Sciences. Surveying would be his life's work— although for some reason he also passed the University of Copenhagen's examination in Roman law in 1778—and by 1798 he had risen to a supervisory position. He "retired" in 1805 but continued to work for several more years before rheumatism, a bad ailment for anyone but particularly so for a surveyor, forced him to really stop. He was highly regarded as a surveyor, receiving a silver medal from the Royal Danish Academy of Sciences for the mapping he did for the French government.

Still, respected as he was as a surveyor, he was not the most obvious person to present a paper (on March 10, 1797, just before his fifty-second birthday) titled "On the Analytic Representation of Direction: An Attempt" to the Royal Danish Academy of Sciences. Wessel is known to have been aided in the writing of his paper by the president of the science section of the Academy, whose support certainly did not hurt, but the intellectual content was all Wessel's. Its quality and merit were judged to be so high that it was the first paper to be accepted for publication—in the Academy's *Memoires* of 1799—by an author not a member of the Academy.

Written in Danish and published in a journal not read by many outside of Denmark, Wessel's brilliant paper was fated to have no impact. It wasn't until 1895 that it was rediscovered and Wessel was at long last recognized as the pioneer he was.[2] Although others, within a few years after Wessel's paper was published, traveled much the same path as he did and it was those others who were read, it was Wessel who walked that path first. Let us examine what he did.

Unlike Wallis' tortuous geometric constructions, figure 3.1 shows the simplicity of Wessel's interpretation of complex numbers. For both him and us modern folk a complex number is either the point $a + ib$ in the so-called *complex plane* (a and b are two real numbers) or the directed radius vector from the origin to that point. Written in this fashion, the complex number is said to be in *rectangular* or *Cartesian* form. An enormously useful alternative is the *polar* form, written in terms of the length of the radius vector and the polar angle, which is the angle measured counterclockwise from the x-axis to the radius vector. That is, if $\theta = \tan^{-1}(b/a)$, then $a + ib = \sqrt{a^2 + b^2}\{\cos(\theta) + i\sin(\theta)\}$. The value of $\sqrt{a^2 + b^2}$, the length of the radius vector, is called the *modulus* of the complex number $a + ib$. The value of the polar angle $\tan^{-1}(b/a)$ is called the *argument* of $a + ib$ and written as $\arg(a + ib)$. More compactly, we can write all this in the form

$$a + ib = \sqrt{a^2 + b^2} \angle \tan^{-1}\left(\frac{b}{a}\right).$$

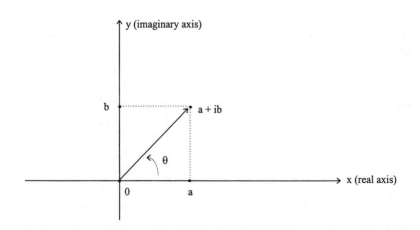

Figure 3.1. Wessel's geometry of complex numbers.

The $\angle$ notation for complex numbers is most commonly used by electrical engineers, but mathematicians often find it quite useful as well. I'll make good use of it in the pages that follow. There is, however, one caveat concerning the polar form of representing complex numbers that is most important to keep in mind. A common error made by students who are first learning about the polar form is a failure to appreciate that the tangent function is periodic with period 180°, *not* 360°. That is, the tangent function goes through its complete interval of values ($-\infty$ to ∞) as the polar angle θ varies from $-90°$ to $90°$. Or, if we express angles in units of radians (one radian $= 180°/\pi = 57.296°$), then the tangent function goes through its complete interval of values as the polar angle varies from $-\pi/2$ to $\pi/2$ radians. This means that blindly plugging values of a and b into $\theta = \tan^{-1}(b/a)$ may lead to mistakes. So, to be precise, let us define

$$\theta = \tan^{-1}\left(\frac{b}{a}\right) \text{ when } a > 0,$$

i.e., the complex number is in either the first or the fourth quadrant, and

$$\theta = 180° + \tan^{-1}\left(\frac{b}{a}\right) \text{ when } a < 0,$$

i.e., the complex number is in either the second or the third quadrant.

Recall that the quadrants of the plane are numbered counterclockwise from the first, which is the quadrant where a and b are both positive. When you use a hand-held calculator's arctangent button, or more likely the SHIFT or INV and then the TAN buttons to compute angles, the machine assumes $a > 0$ (if $b/a < 0$ the machine assumes it is because of b) and so it always returns a result between $-90°$ and $90°$. This is called the *principal value* of the arctangent. If, in fact, $a < 0$, then *you* must make the appropriate adjustment as given above.

How was Wessel led to the now standard way of representing complex numbers? Wessel began his paper by describing what today is called vector addition. That is, if we have two directed line segments both lying along the x-axis (but perhaps in opposite directions), then we add them by positioning the starting point of one at the terminal point of the other, and the sum is the net resulting directed line segment extending from the initial point of the first segment to the terminal point of the second. Wessel said the sum of two nonparallel segments should obey the same rule, and this procedure is shown in figure 3.2. So far there is nothing new here, as Wallis had expressed quite similar ideas on how to add directed line segments. Wessel's original contribution was to see how to multiply such segments.

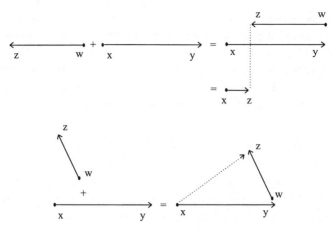

Figure 3.2. Vector addition.

Wessel discovered how to multiply line segments by making a clever generalization from the behavior of real numbers. He observed that the product of two numbers (say, 3 and -2, with a product of -6) has the same ratio to each factor as the other factor has to 1. That is, $-6/3 = -2 = -2/1$, and $-6/-2 = 3 = 3/1$. So, assuming there exists a unit directed line segment, Wessel argued that the product of two directed line segments should have two properties. First, and immediately analogous to real numbers, the length of the product should be the product of the lengths of the individual line segments.

But what of the direction of the product? This second property is Wessel's seminal contribution: by analogy with all that has gone before, he said the line segment product should differ in direction from each line segment factor by the same angular amount as the other line segment factor differs in direction when compared to the unit directed line segment. Thus, suppose the unit directed line segment points from left to right at angle $0°$, i.e., along the positive x-axis. Then, if we have two line segments to multiply together, one at angle θ and the other at angle α, the product angle must be the sum $\theta + \alpha$ because $\theta + \alpha$ differs from θ by α (the angle by which the α-segment differs from the unit segment), and $\theta + \alpha$ differs from α by θ (the angle by which the θ-segment differs from the unit segment).

This is perhaps so painfully elementary, or even obvious, today, that it seems almost childish to write it out in so many words, but don't be deceived. If you think it "obvious" you are confusing something known for a long time, perhaps since high school, with something everybody is born knowing. Babies are born knowing how to cry, but they are *not* born knowing how to multiply

51

directed line segments together. That is something that had to be either discovered or invented, depending on how you view the evolution of mathematics, and Wessel was the first to see how to do it.

With just these few brilliant insights by Wessel we can do some extraordinary calculations. Let me show you just three. First, what is $(0.3 + i2.6)^{17}$? This looks pretty horrible at first, but observe that

$$0.3 + i2.6 = \sqrt{(0.3)^2 + (2.6)^2} \; \angle \tan^{-1}\left(\frac{2.6}{0.3}\right) = 2.6172505 \; \angle \; 83.418055°.$$

Since $(0.3 + i2.6)^{17}$ means $0.3 + i2.6$ multiplied by itself seventeen times, then Wessel's rules tell us to raise the magnitude or modulus to the seventeenth power, and to multiply the polar angle or argument by seventeen. That is,

$$(0.3 + i2.6)^{17} = (2.6172505)^{17} \; \angle \; (83.418055° \times 17) = 12{,}687{,}322 \; \angle \; 1{,}418.1061°.$$

This is a complex number in the fourth quadrant, which you can see by simply subtracting 360° chunks of polar angle—with each such chunk representing one complete revolution about the origin—until you arrive at an angle less than 360°. Thus,

$$(0.3 + i\,2.6)^{17} = 12{,}687{,}322 \; \angle \; 338.1061°$$
$$= 12{,}687{,}322 \; \angle \; -21.893915°$$
$$= 12{,}687{,}322 \; \{\cos(-21.893915°) + i\,\sin(-21.893915°)\}$$
$$= 11{,}772{,}300 - i\,4{,}730{,}800.$$

Before Wessel this calculation would have required multiplying $0.3 + i2.6$ by itself seventeen times, and the details would have driven most people crazy.

Here is another even more wonderful computation based on Wessel's idea of adding the angles. Consider the product $(2 + i)(3 + i) = 5 + i5$. Using the radian as our unit of angle, the angle of the product is $\tan^{-1}(5/5) = \tan^{-1}(1) = \pi/4$. The angles of the two factors on the left are, similarly, $\tan^{-1}(1/2)$ and $\tan^{-1}(1/3)$. Thus, we immediately have

$$\tan^{-1}\left(\frac{1}{2}\right) + \tan^{-1}\left(\frac{1}{3}\right) = \frac{\pi}{4},$$

a result you can easily verify on a calculator set to its radian, not degree, mode. I defy you to find a more direct derivation of this formula. In the same way, multiply out $(5 + i)^4(-239 + i)$ and see if you can demonstrate that

$$4 \tan^{-1}\left(\frac{1}{5}\right) - \tan^{-1}\left(\frac{1}{239}\right) = \frac{\pi}{4}.$$

This is a famous formula in the history of π, one you will see again in chapter 6. After you do the multiplication you should find that the product is a number in the third quadrant, with angle $5\pi/4$ radians, not $\pi/4$ radians. This may puzzle you at first but, if so, probably not after you observe that the factor $(-239 + i)$ is in the second quadrant and so has angle $\{\pi - \tan^{-1}(1/239)\}$ radians. After you collect terms you will see that the three terms on the left-hand side of the identity sum to $-3\frac{3}{4}\pi$ radians, which is the angle of a directed line segment in the first quadrant with an angle of $\pi/4$ radians (remember, the minus sign says to rotate clockwise from the positive real axis).

And finally, if you multiply out $(p + q + i)(p^2 + pq + 1 + iq)$, where p and q are any pair of real numbers, you should now easily be able to derive the well-known identity

$$\tan^{-1}\{1/(p + q)\} + \tan^{-1}\{q/(p^2 + pq + 1)\} = \tan^{-1}(1/p).$$

Without complex numbers and Wessel's idea of adding the angles, deriving this identity would be a much more difficult problem.

Ever since Wessel, then, multiplying two directed line segments together has meant the two-step operation of multiplying the two lengths, with length always taken to be a positive value, and adding the two direction angles. These two operations determine the length and direction angle of the product, and it is this definition of a product that gives us the explanation for what $\sqrt{-1}$ means geometrically. That is, suppose that there *is* a directed line segment that represents $\sqrt{-1}$, and that its length is l and its direction angle θ. Mathematically, then, we have $\sqrt{-1} = l \angle \theta$. Multiplying this statement by itself, i.e., squaring both sides, we have $-1 = l^2 \angle 2\theta$ or, as $-1 = 1 \angle 180°$, then $l^2 \angle 2\theta = 1 \angle 180°$. Thus, $l^2 = 1$ and $2\theta = 180°$, and so $l = 1$ and $\theta = 90°$. This says $\sqrt{-1}$ is the directed line segment of length one pointing straight up along the vertical axis or, at long last,

$$\boxed{i = \sqrt{-1} = 1 \angle 90°.}$$

This is so important a statement that it is the only mathematical expression in the entire book that I have enclosed.

Historians generally credit Wessel with being the first to associate an axis perpendicular to the real axis with the axis of imaginaries. However, there are indications that this idea, just before Wessel, was one whose time was ripe. There is a throwaway line, for example, in an 1847 book by the great French mathematical physicist Augustin-Louis Cauchy that as early as 1786 one

Henri Dominique Truel ("a modest scholar," wrote Cauchy) had represented imaginary values by an axis perpendicular to the horizontal real axis. Nothing is known of Truel, however, and it appears he never published his results, whatever they may have been. And there are hints in the writings of Gauss that he was onto the same idea as early as 1796, but he too did not publish at that time. It was Wessel who was first in putting his ideas forward in a public forum.

How beautifully simple is Wessel's idea. Multiplying by $\sqrt{-1}$ is, geometrically, simply a rotation by 90° in the counterclockwise (CCW) sense. In figure 3.3, for example, the vector representing the complex number $a + ib$ is drawn in the first quadrant of the complex plane ($a > 0$, $b > 0$). Multiplying by i gives a 90° CCW rotation to the vector, rotating it into the second quadrant to $i(a + ib) = -b + ia$. Because of this property $\sqrt{-1}$ is often said to be the *rotation operator,* in addition to being an imaginary number.

As one historian of mathematics has observed,[3] the elegance and sheer wonderful simplicity of this interpretation suggests "that there is no occasion for anyone to muddle himself into a state of mystic wonderment over . . . the grossly misnamed 'imaginaries.'" This is not to say, however, that this geo-

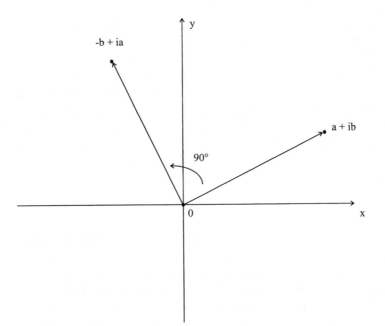

Figure 3.3. $\sqrt{-1}$ as the *rotation operator* in the complex plane.

metric interpretation wasn't a huge leap forward in human understanding. Indeed, it is only the start of a tidal wave of elegant calculations.

For example, figure 3.4 shows the unit circle and two arbitrary radius vectors, one at angle θ and one at angle α. Since each vector has unit length, their product will also be of unit length, at angle $\alpha + \theta$ (as shown). Writing all three radius vectors out mathematically, we have

$$1 \angle \theta = \cos(\theta) + i \sin(\theta),$$

$$1 \angle \alpha = \cos(\alpha) + \sin(\alpha),$$

$$1 \angle (\alpha + \theta) = \cos(\alpha + \theta) + i \sin(\alpha + \theta),$$

and since $(1 \angle \theta)(1 \angle \alpha) = 1 \angle (\alpha + \theta)$, we must have

$$\{\cos(\theta) + i \sin(\theta)\} \{\cos(\alpha) + i \sin(\alpha)\} = \cos(\alpha + \theta) + i \sin(\alpha + \theta).$$

Expanding the left-hand side gives

$$\{\cos(\theta)\cos(\alpha) - \sin(\theta)\sin(\alpha)\} + i \{\sin(\theta)\cos(\alpha) + \cos(\theta)\sin(\alpha)\} = \cos(\alpha + \theta) + i \sin(\alpha + \theta).$$

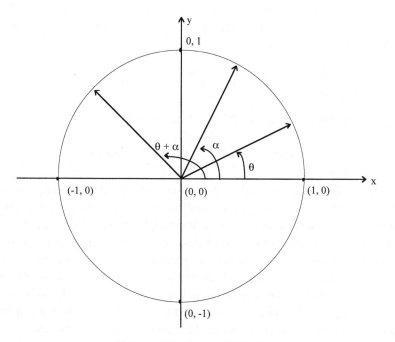

Figure 3.4. Vector multiplication.

Two complex numbers are equal only if their real and imaginary parts are separately equal, and so we immediately have the two trigonometric identities

$$\cos(\alpha + \theta) = \cos(\theta)\cos(\alpha) - \sin(\theta)\sin(\alpha),$$

$$\sin(\alpha + \theta) = \sin(\theta)\cos(\alpha) + \cos(\theta)\sin(\alpha).$$

For the special case of $\alpha = \theta$ these expressions reduce to $\cos(2\alpha) = \cos^2(\alpha) - \sin^2(\alpha)$ and $\sin(2\alpha) = 2\sin(\alpha)\cos(\alpha)$. Such identities had been known long before Wessel, of course; for example, they can be found in the book *Almagest* (the greatest) by Ptolemy of Alexandria, written in the second century A.D., but until Wessel's new geometry of complex numbers they had never before been so easily derived. By deriving these identities in this way I have deviated from Wessel's actual presentation. He used the identities to do things in essentially the reverse order from my presentation, but I think showing the development the way I have is much more dramatic. The intellectual content is the same, either way. These two identities are useful, for example, in deriving the expression for $\tan(\alpha - \beta)$ used in section 2.1. Simply write $\tan(\alpha - \beta) = \sin(\alpha - \beta)/\cos(\alpha - \beta)$, use the two above identities to expand the sine and cosine, and divide through by $\cos(\alpha)\cos(\beta)$.

With his wonderful deduction of the geometry of $\sqrt{-1}$ there was now no stopping Wessel with even more exotic calculations. For example, if you start with a unit radius vector of direction angle θ/m, where m is an integer, then it follows immediately that

$$\left\{ 1 \angle \frac{\theta}{m} \right\}^m = \left\{ \cos\left(\frac{\theta}{m}\right) + i\, \sin\left(\frac{\theta}{m}\right) \right\}^m = 1 \angle \theta = \cos(\theta) + i\, \sin(\theta).$$

Or, turning this statement around by taking the mth root,

$$\{\cos(\theta) + i\, \sin(\theta)\}^{1/m} = \cos\left(\frac{\theta}{m}\right) + i\, \sin\left(\frac{\theta}{m}\right).$$

This result was not original with Wessel (although this elegantly simple derivation of it was), and it is commonly known as "De Moivre's theorem," after the French-born mathematician Abraham De Moivre (1667–1754). De Moivre, a Protestant, left Catholic France at age eighteen to seek religious freedom in London, where he became a friend of Isaac Newton. In a 1698 paper published in the *Philosophical Transactions* of the Royal Society, he mentions that Newton knew of an equivalent expression of De Moivre's theorem as early as 1676, which Newton used to calculate the cube roots of the complex numbers that come out of the Cardan formula for the irreducible case. De Moivre probably learned of this technique from Newton, and used it in earning his livelihood as a "solver of mathematical problems." De Moivre was a man of great talent—particularly skilled in probability theory applied to gambling, he wrote *The Doctrine of Chances* in 1718 and discovered the ubiquitous normal or "bell-shaped" curve now known by Gauss' name. Tradition has it that Newton himself would often reply to mathematical questions by saying "Ask Mr. De Moivre, he knows more about it than I do." It is clear from De Moivre's writings that he did, in fact, know and use the above result, but he never actually wrote it out explicitly—that was done by Euler in 1748, who arrived at it by entirely different means, which are discussed in chapter 6.

De Moivre's theorem allowed Wessel to calculate any integer root of a complex number. In his paper, for example, Wessel hints at the power of his results by stating

$$\sqrt[3]{4\sqrt{3}+4\sqrt{-1}} = 2 \angle 10°.$$

This is, indeed, correct, but he doesn't provide any of the details on how he arrived at it. Here's how he probably reasoned. The complex number under the cube root sign is, in polar form,

$$\sqrt{(4\sqrt{3})^2 + (4)^2} \angle \tan^{-1}\left(\frac{4}{4\sqrt{3}}\right) = \sqrt{48+16} \angle \tan^{-1}\left(\frac{1}{\sqrt{3}}\right) = 8 \angle 30°.$$

Thus,

$$\sqrt[3]{4\sqrt{3}+4\sqrt{-1}} = (8 \angle 30°)^{1/3} = 8^{1/3} \angle \frac{30°}{3} = 2 \angle 10°,$$

which is, when written out in Cartesian form,

$$2\cos(10°) + i\,2\sin(10°) = 1.969615506 + i\,0.347296355.$$

That this result is correct can be verified simply by cubing it, a tedious task made easy with a hand-held calculator that can do complex number arithmetic, and observing that the result is, indeed, $6.928203230 + i4 = 4\sqrt{3} + i4$. Wessel was well aware that, just as there are two square roots of any number, there are m mth roots. So there are three cube roots of $4\sqrt{3} + i4$. Wessel also knew that these roots are separated by equal angles, i.e., if one root is $2 \angle 10°$ then the other two are $2 \angle 130°$ and $2 \angle 250°$. To see this, suppose the complex number for which we are calculating the nth roots has angle θ. Then, obviously, one of the roots will have angle θ/n because when we raise that root to the nth power we must get the original number back, at angle θ. This is so since angles add during multiplication and we have $\theta/n \times n = \theta$. If another possible angle for a root is α, then $n\alpha = \theta + k \cdot 360°$, where k is some integer, i.e., adding an integer number of complete 360° rotations to θ makes no difference as it returns us to θ. Thus $\alpha = \theta/n + k \cdot 360°/n$. When $k = 0$ we get $\alpha = \theta/n$, the angle of the "obvious" root. And for $k = 1, 2, \ldots, n - 1$ we get $n - 1$ more, different angles, for a total of n uniformly spaced angles. Using other integer values for k that are either negative or greater than $n - 1$ simply repeats roots given by values of k equal to the first n non-negative integers, e.g., $k = n$ repeats the root given by $k = 0$.

As another example of this sort of calculation, recall from chapter 1 the Cardan formula solution to the irreducible cubic considered by Bombelli,

$$x = \sqrt[3]{2 + \sqrt{-121}} + \sqrt[3]{2 - \sqrt{-121}}.$$

There I went through a fairly complicated argument to show that

$$\sqrt[3]{2 + \sqrt{-121}} = 2 + \sqrt{-1},$$

$$\sqrt[3]{2 - \sqrt{-121}} = 2 - \sqrt{-1},$$

and so $x = 4$. Using De Moivre's formula, however, these results are easy to calculate. Thus, the complex number $2 + \sqrt{-121} = 2 + i11$ is, in polar form,

$$\sqrt{(2)^2 + (11)^2} \angle \tan^{-1}\left(\frac{11}{2}\right) = \sqrt{125} \angle 79.69515353°.$$

So,

$$\sqrt[3]{2 + \sqrt{-121}} = \left(\sqrt{125}\right)^{1/3} \angle \frac{79.69515353°}{3}$$

$$= 2.236067977 \angle 26.56505117°$$

$$= 2.236067977\{\cos(26.56505117° + i\ \sin(26.56505117°)\}$$

$$= 2 + i$$

Since there are two other cube roots of $2 + \sqrt{-121}$ (and similarly for $2 - \sqrt{-121}$), there are a total of three values of x that give the three roots of the cubic equation. All will be real and, of course, equal to the roots calculated in section 1.6 using Viète's trigonometric solution.

As one final example, consider the problem of finding the nth roots of unity; that is, of finding all the solutions to $z^n = 1$, the so-called *cyclotomic* equation. The name refers to the close association of the equation with the construction of regular n-gons inscribed in circles, a problem I'll return to in the next chapter. The equation can be rewritten as $z^n - 1 = 0$, and you can easily verify by long division that the left-hand side can be factored as

$$(z - 1)(z^{n-1} + z^{n-2} + \cdots + z + 1) = 0.$$

THE PUZZLES START TO CLEAR

One solution is the obvious root of 1 from the left-hand factor, and so the other $n-1$ roots must be the roots of the right-hand factor. That is, the solutions to the rather formidable-looking equation

$$z^{n-1} + z^{n-2} + \cdots + z + 1 = 0$$

are simply those given by De Moivre's formula, i.e.,

$$z = \cos\left(k\frac{2\pi}{n}\right) + i\sin\left(k\frac{2\pi}{n}\right), \; k = 1, 2, \ldots, n-1.$$

The case of $k = 0$ gives, of course, the $z = 1$ solution that comes from the $(z - 1)$ factor.

For the case of $n = 5$, for example, the four roots (other than the obvious $z = 1$) to the quintic cyclotomic $z^5 - 1 = 0$ are given by

$$k = 1; \cos\left(\frac{2\pi}{5}\right) + i\,\sin\left(\frac{2\pi}{5}\right) = 0.309017 + i0.9510565,$$

$$k = 2; \cos\left(\frac{4\pi}{5}\right) + i\,\sin\left(\frac{4\pi}{5}\right) = -0.809017 + i0.5877853,$$

$$k = 3; \cos\left(\frac{6\pi}{5}\right) + i\,\sin\left(\frac{6\pi}{5}\right) = -0.809017 - i0.5877853,$$

$$k = 4; \cos\left(\frac{8\pi}{5}\right) + i\,\sin\left(\frac{8\pi}{5}\right) = 0.309017 - i0.9510565,$$

Now, to help you feel more comfortable with this result, as well as to demonstrate dramatically how easily De Moivre's theorem generates solutions to the cyclotomic equation, I will show you a completely different, *algebraic* solution for the $n = 5$ case. As the Italian-born Joseph Louis Lagrange (1736–1813) did in 1771, I will first remove the obvious $z - 1$ factor to get the reduced quartic equation

$$z^4 + z^3 + z^2 + z + 1 = 0.$$

This can be written as

$$(z^2 + z^{-2}) + (z + z^{-1}) + 1 = 0,$$

which, be defining $u = z + z^{-1}$ and noticing that $u^2 = z^2 + z^{-2} + 2$, becomes

$$u^2 + u - 1 = 0.$$

This quadratic is quickly solved to give

$$u = \frac{-1 \pm \sqrt{5}}{2} = 0.618034 \text{ and } -1.618034.$$

Next, from the definition of u we have $z^2 - uz + 1 = 0$, another quadratic also quickly solved to give

$$z = \frac{u \pm \sqrt{u^2 - 4}}{2}$$

Substitution of the above values for u gives us four values of z that are exactly what De Moivre's theorem gave. Complex number geometry and De Moivre's theorem, and ordinary algebra, are in total agreement here, and that should increase your confidence in the truth of the claim that there is nothing at all imaginary about complex numbers. De Moivre's theorem would in fact give us the solutions to $z^{97} - 1 = 0$ just as quickly as it did for $n = 5$ (or for any other integer n), while Lagrange's clever algebraic substitution that works so well for $n = 5$ does *not* work in the general case.

A slightly more sophisticated analysis than that provided here would show that De Moivre's theorem is true even when m is not restricted to positive integers. An especially interesting special case, for example, is for $m = -1$, which says

$$\{\cos(\theta) + i\sin(\theta)\}^{-1} = \frac{1}{\cos(\theta) + i\sin(\theta)} = \cos(-\theta) + i\sin(-\theta)$$
$$= \cos(\theta) - i\sin(\theta).$$

That is, on the circumference of the unit circle the reciprocal of any complex number is equal to the conjugate of that number. This result becomes self-evident, in retrospect, by cross-multiplying to get the well-known identity $1 = \{\cos(\theta) + i\sin(\theta)\}\{\cos(\theta) - i\sin(\theta)\} = \cos^2(\theta) + \sin^2(\theta)$.

3.2 DERIVING TRIGONOMETRIC IDENTITIES FROM DE MOIVRE'S THEOREM

De Moivre's theorem is, in fact, a machine for generating trigonometric identities. If, for example, we set $m = 3$ in De Moivre's theorem, then we get

$$\left\{\cos\left(\frac{1}{3}\theta\right) + i\sin\left(\frac{1}{3}\theta\right)\right\}^3 = \cos(\theta) + i\sin(\theta).$$

Expanding the left-hand side gives

$$\cos^3\left(\frac{1}{3}\theta\right) - 3\sin^2\left(\frac{1}{3}\theta\right)\cos\left(\frac{1}{3}\theta\right) + i\left[3\cos^2\left(\frac{1}{3}\theta\right)\sin\left(\frac{1}{3}\theta\right) - \sin^3\left(\frac{1}{3}\theta\right)\right],$$

which, when the real parts of this and of the right-hand side of the first identity are set equal, with $\theta = 3\alpha$, results in

$$\cos(3\alpha) = \cos^3(\alpha) - 3\sin^2(\alpha)\cos(\alpha).$$

Now, since $\sin^2(\alpha) + \cos^2(\alpha) = 1$, an identity derived at the end of the last section, we can write

$$\cos(3\alpha) = 4\cos^3(\alpha) - 3\cos(\alpha)$$

or

$$\cos^3(\alpha) = \frac{3}{4}\cos(\alpha) + \frac{1}{4}\cos(3\alpha).$$

This is the trigonometric identity used by Viète in solving the irreducible case of the Cardan formula. Using other values of m in De Moivre's theorem, and separately equating real and imaginary parts of the result, allows the derivation of literally an endless number of trigonometric identities.

There is another, ingenious way to use De Moivre's theorem to derive trigonometric identities. If we let z denote a point on the unit circle in the complex plane, then we can write z in Cartesian form as $z = \cos(\theta) + i\sin(\theta)$, where θ is the angle of the radius vector directed from the origin to z. As shown in the previous section, $1/z = z^{-1} = \cos(\theta) - i\sin(\theta)$. From De Moivre's theorem we also have $z^n = \cos(n\theta) + i\sin(n\theta)$, and $z^{-n} = \cos(n\theta) - i\sin(n\theta)$. All these statements together allow us to write

$$z + z^{-1} = 2\cos(\theta) \text{ and } z^n + z^{-n} = 2\cos(n\theta).$$

Now, for a specific example of what we can do with these two results. Suppose we write

$$(z + z^{-1})^6 = 2^6\cos^6(\theta).$$

We could also multiply $(z + z^{-1})^6$ out to get

$$z^6 + 6z^4 + 15z^2 + 20 + 15z^{-2} + 6z^{-4} + z^{-6} = (z^6 + z^{-6}) + 6(z^4 + z^{-4}) + 15(z^2 + z^{-2}) + 20.$$

We don't have to actually multiply this out term by term. It is much easier to use the binomial theorem, which says

$$(x + y)^n = \sum_{k=0}^{n} \binom{n}{k} x^{n-k} y^k$$

where the binomial coefficient

61

$$\binom{n}{k} = \frac{n!}{k!(n-k)!},$$

(Remember that $0! = 1$, but if you have forgotten I'll derive it for you in chapter 6.) It is understood that

$$\binom{n}{k} = 0,$$

if $k > n$. Now let $x = z$ and $y = z^{-1}$ and you can write down the terms of $(z + z^{-1})^6$ as fast as you can evaluate factorials. Now, the above result for $(z + z^{-1})^6$ is equal to the following [where I have successively taken $n = 6, 4,$ and 2 in the relation $z^n + z^{-n} = 2\cos(n\theta)$]:

$$2\cos(6\theta) + 12\cos(4\theta) + 30\cos(2\theta) + 20.$$

But this is equal to $2^6\cos^6(\theta)$, and so we have the exotic identity

$$\cos^6(\theta) = \frac{1}{32}\cos(6\theta) + \frac{3}{16}\cos(4\theta) + \frac{15}{32}\cos(2\theta) + \frac{5}{16}.$$

The presence of the constant term, $\frac{5}{16}$, is due to the fact that $\cos^6(\theta)$ is never negative, i.e., electrical engineers would say the average or *dc value* of $\cos^6(\theta)$ is $\frac{5}{16}$. The dc value of $\cos^{15}(\theta)$, on the other hand, is zero because it is symmetrical about the horizontal axis, i.e., it is negative and positive equally. It would be no more difficult to derive a similar result for, say, $\cos^{15}(\theta)$.

Now, to demonstrate to yourself how much power this application of $\sqrt{-1}$ has given you, see if you can verify the following amazing identity: $\cos(11\theta)$ = $1024\cos^{11}(\theta) - 2{,}816\cos^9(\theta) + 2{,}816\cos^7(\theta) - 1{,}232\cos^5(\theta) + 220\cos^3(\theta) - 11\cos(\theta)$. Can you also derive the series for $\sin(11\theta)$ in terms of powers of $\sin(\theta)$? This *is* a bit of work, but you should know that in 1593 the Belgium-born mathematician Adrian van Roomer (1561–1615) worked out such a power series for $\sin(45\theta)$ while constructing a famous challenge problem—that of solving a particular forty-fifth-degree equation. Viète astonished van Roomer by using his vast knowledge of trigonometry to solve the equation in just one day. Now *that* is a lot harder than working out $\sin(11\theta)$, don't you think?

De Moivre's formula appears to be able to do everything we might want for sines and cosines, but what about the other trigonometric functions? Complex numbers can handle such questions, too. In chapter 6, for example, you'll see how one of De Moivre's contemporaries developed an ingenious, calculus-based use of $\sqrt{-1}$ to derive a general formula for $\tan(n\theta)$ in terms of powers of $\tan(\theta)$.

As a dramatic example of De Moivre's theorem, let me show you how it can be used to derive Viète's infinite product formula for π that I mentioned back in note 6 of chapter 1. Starting with De Moivre's theorem for $m = 2$, we have

$$\{\cos(\theta) + i\sin(\theta)\}^{1/2} = \cos\left(\frac{1}{2}\theta\right) + i\sin\left(\frac{1}{2}\theta\right).$$

Squaring both sides, we get

$$\cos(\theta) + i\sin(\theta) = \cos^2\left(\frac{1}{2}\theta\right) - \sin^2\left(\frac{1}{2}\theta\right) + i\,2\sin\left(\frac{1}{2}\theta\right)\cos\left(\frac{1}{2}\theta\right)$$

which, when the real parts are equated, gives $\cos(\theta) = \cos^2(\frac{1}{2}\theta) - \sin^2(\frac{1}{2}\theta)$, a result derived in section 3.1 as a special case of the identity for $\cos(\alpha + \theta)$. Using the identity $\cos^2(\frac{1}{2}\theta) + \sin^2(\frac{1}{2}\theta) = 1$ then gives us the so-called half-angle formulas

$$\cos\left(\frac{1}{2}\theta\right) = \sqrt{\frac{1 + \cos(\theta)}{2}}$$

and

$$\sin\left(\frac{1}{2}\theta\right) = \sqrt{\frac{1 - \cos(\theta)}{2}}.$$

If we equate the imaginary parts, however, we get

$$\sin(\theta) = 2\sin\left(\frac{1}{2}\theta\right)\cos\left(\frac{1}{2}\theta\right).$$

This identity can be applied to itself to write

$$\sin(\theta) = 2\left\{2\sin\left(\frac{1}{4}\theta\right)\cos\left(\frac{1}{4}\theta\right)\right\}\cos\left(\frac{1}{2}\theta\right).$$

In fact, we can continue to do this, and if we do it n times (where $n = 1$ is the original identity) then

$$\sin(\theta) = 2^n\cos\left(\frac{1}{2}\theta\right)\cos\left(\frac{1}{4}\theta\right)\cdots\cos\left(\frac{1}{2^n}\theta\right)\sin\left(\frac{1}{2^n}\theta\right).$$

Dividing through by θ gives

$$\frac{\sin(\theta)}{\theta} = \cos\left(\frac{1}{2}\theta\right)\cos\left(\frac{1}{4}\theta\right)\cdots\cos\left(\frac{1}{2^n}\theta\right)\frac{\sin\left(\frac{1}{2^n}\theta\right)}{\left(\frac{1}{2^n}\theta\right)}$$

63

If we then let n become arbitrarily large, and θ is expressed in units of radians, then the last factor converges to 1 (if you don't recall why this is so, don't worry—I'll derive it for you in chapter 6). So

$$\frac{\sin(\theta)}{\theta} = \cos\left(\frac{1}{2}\theta\right)\cos\left(\frac{1}{4}\theta\right)\cos\left(\frac{1}{8}\theta\right)\cdots,$$

where the right-hand side continues on for an infinity of factors. Now, for $\theta = \pi/2$, this reduces to Viète's formula

$$\frac{2}{\pi} = \cos\left(\frac{\pi}{4}\right)\cos\left(\frac{\pi}{8}\right)\cos\left(\frac{\pi}{16}\right)\cdots.$$

Using Π as the product symbol, this is more compactly written as

$$\prod_{k=2}^{\infty}\cos\left(\frac{\pi}{2^k}\right) = \frac{2}{\pi}.$$

Since $\cos(\pi/4) = \sqrt{2}/2$, applying the half-angle formula for $\cos(\frac{1}{2}\theta)$ gives the elegant-appearing equivalent expression

$$\frac{2}{\pi} = \frac{\sqrt{2}}{2}\cdot\frac{\sqrt{2+\sqrt{2}}}{2}\cdot\frac{\sqrt{2+\sqrt{2+\sqrt{2}}}}{2}\cdots.$$

We can use the results of this section and the previous one to solve equations that would otherwise be very difficult to attack. For example, what are all the solutions to $(z + 1)^n = z^n$, where n is a positive integer? Since this is a polynomial equation of degree $n - 1$ (notice that the z^n terms on both sides cancel) we expect that there are $n - 1$ solutions. As $z = 0$ is obviously *not* a solution, we can divide through by z and say that the original equation is equivalent to

$$\left(\frac{z+1}{z}\right)^n = 1,$$

and so from section 3.1 we immediately have

$$\frac{z+1}{z} = \cos\left(k\frac{2\pi}{n}\right) + i\sin\left(k\frac{2\pi}{n}\right), \quad k = 0, 1, 2, \ldots, n-1.$$

Cross-multiplying and collecting terms gives

$$z\left\{1 - \cos\left(k\frac{2\pi}{n}\right) - i\sin\left(k\frac{2\pi}{n}\right)\right\} = -1.$$

From one of the above half-angle formulas we know that $1 - \cos(k2\pi/n) = 2\sin^2(k\pi/n)$. From section 3.1 we also know that $\sin(k2\pi/n) = 2\sin(k\pi/n)\cos(k\pi/n)$. Thus,

$$z\left\{2\sin^2\left(k\frac{\pi}{n}\right) - i\,2\sin\left(k\frac{\pi}{n}\right)\cos\left(k\frac{\pi}{n}\right)\right\} = -1.$$

Since $-i^2 = 1$ then this last statement can be written as

$$-i2z\sin\left(k\frac{\pi}{n}\right)\left\{\cos\left(k\frac{\pi}{n}\right) + i\sin\left(k\frac{\pi}{n}\right)\right\} = -1$$

or, solving for z,

$$z = \frac{1}{i2\sin\left(k\dfrac{\pi}{n}\right)\left\{\cos\left(k\dfrac{\pi}{n}\right) + i\sin\left(k\dfrac{\pi}{n}\right)\right\}}.$$

There is an important proviso here, however—we must exclude the $k = 0$ case to avoid dividing by zero, i.e., to avoid the $\sin(k\pi/n) = 0$ case. This gives us our $n - 1$ solutions, then, as now $k = 1, 2, \ldots, n - 1$. We can put the solutions in much more elegant form by using the last result from the previous section, i.e.,

$$\frac{1}{\cos\left(k\dfrac{\pi}{n}\right) + i\sin\left(k\dfrac{\pi}{n}\right)} = \cos\left(k\frac{\pi}{n}\right) - i\sin\left(k\frac{\pi}{n}\right).$$

Thus,

$$z = \frac{\cos\left(k\dfrac{\pi}{n}\right) - i\sin\left(k\dfrac{\pi}{n}\right)}{i2\sin\left(k\dfrac{\pi}{n}\right)} = -\frac{1}{2} - i\frac{1}{2}\cot\left(k\frac{\pi}{n}\right), \quad k = 1, 2, \ldots, n-1,$$

which gives us the rather surprising (I think) result that all of the solutions to $(z + 1)^n = z^n$ each lie on the vertical line that intersects the real axis at $x = -\frac{1}{2}$. Without De Moivre's theorem this would be a very much more difficult problem.

3.3 COMPLEX NUMBERS AND EXPONENTIALS

The introduction of the *complex plane* into mathematics by Wessel dramatically expanded the concept of *number*. Before Wessel all known numbers were real, confined to the one-dimensional *x*-axis, the so-called real axis. After Wessel, the domain of all possible numbers expanded to the two-dimensional plane, infinite in all directions, not just to the left and right. The discovery of complex numbers was the last in a sequence of discoveries that gradually filled in the set of all numbers, starting with the positive integers (finger counting) and then expanding to include the positive rationals and irrational reals, negatives, and then finally the complex.

Although I will not prove it here, the complex numbers are *complete* under the usual arithmetic operations. That means we will always get another complex number as the result of doing addition, subtraction, multiplication, division, taking any root, etc., on other complex numbers. When we try to take the square root of -1 (a real number), for example, we suddenly leave the real numbers, and so the reals are not complete with respect to the square root operation. We don't have to be concerned that something like that will happen with the complex numbers, however, and we won't have to invent even more exotic numbers (the "really complex"!) Complex numbers are everything there is in the two-dimensional plane.

Now, just one last comment on Wessel's work before we move on. After the development of the brilliant ideas I've discussed so far, Wessel made the statement that there is yet a third way to represent complex numbers, in addition to the Cartesian and the polar forms, a way involving *exponentials*. I'm not going to go into his arguments on this in this book, for two reasons. First, Wessel's presentation on this point is definitely less impressive than is the first part of his paper, depending as it does on a rather questionable amalgamation of the binomial theorem and infinite series—and he actually doesn't do anything with the results he gets. Wessel himself knew things were unclear when he wrote "I shall produce complete proofs for [my claims] at another time, if privileged to do so." He never did. And second, the exponential connection to $\sqrt{-1}$ was done decades before Wessel by others, in far more convincing fashion. I think it is best to see how they did it, and their work appears in chapter 6. As a prelude to that more extensive historical presentation, however, let me next show you how a modern mathematician might develop the exponential interpretation of complex numbers.

Consider two complex numbers on the unit circle, one at angle α and the other at angle θ. Wessel's definition of complex multiplication tells us that

$$\{\cos(\alpha) + i\sin(\alpha)\}\{\cos(\theta) + i\sin(\theta)\} = \cos(\alpha + \theta) + i\sin(\alpha + \theta),$$

as depicted in figure 3.3. De Moivre's theorem tells us that

$$\{\cos(\theta) + i\sin(\theta)\}^n = \cos(n\theta) + i\sin(n\theta).$$

We can write these two statements in compact form with the notation $f(\theta) = \cos(\theta) + i\sin(\theta)$, and similarly for $f(\alpha)$, as follows:

$$f(\alpha)f(\theta) = f(\alpha + \theta),$$
$$f^n(\theta) = f(n\theta).$$

Now, try to think of an actual function, f, that has both of these properties. After a bit, you might think of

$$f(\alpha) = e^{K\alpha},$$
$$f(\theta) = e^{K\theta},$$

where K is a constant. This works because

$$e^{K\alpha} \cdot e^{K\theta} = e^{K(\alpha+\theta)},$$
$$\{e^{K\theta}\}^n = e^{nK\theta}.$$

Thus, we can write $f(\theta) = \cos(\theta) + i\sin(\theta) = e^{K\theta}$, and for the special case of $\theta = 0$ (which reduces to $1 = e^0$) this is obviously true for any value of K. But for the general case of arbitrary θ, K cannot be just anything. To find K, differentiate $f(\theta)$ with respect to θ to get

$$-\sin(\theta) + i\cos(\theta) = Ke^{K\theta} = i\{\cos(\theta) + i\sin(\theta)\} = ie^{K\theta}.$$

Thus, $K = i$ and so $f(\theta) = \cos(\theta) + i\sin(\theta) = 1 \angle \theta = e^{i\theta}$.

This famous result—called "Euler's identity"—will be discussed in much more detail in chapter 6. Just to leave you for now with a few little gems of symbolic dazzle, however, suppose $\theta = \pi$. Then, $-1 = e^{i\pi}$ or $e^{i\pi} + 1 = 0$, a single statement connecting *five* central numbers in mathematics. James Gleick's biography of the great American physicist Richard Feynman reproduces this statement on a page from one of Feynman's youthful notebooks.[4] Dated April 1933, a month before his fifteenth birthday, the page has this identity printed in a bold scrawl and titled "THE MOST REMARKABLE FORMULA IN MATH." This is an exuberant overstatement, but it *is* remarkable as it is a special case of a more general result that allows us to calculate logarithms of negative numbers, and complex ones, too, something students are told in high school algebra can't be done. As an example of how it *can* be done, just write $e^{i\pi} = -1$ and so $\ln(-1) = i\pi$. The logarithm of a negative number is an imaginary number.

If $\theta = \pi/2$ in the expression for $e^{i\theta}$ then $i = e^{i\pi/2}$. If we take the natural logarithm of both sides (assuming such a thing can be done—what, after all, could $\ln(i)$ mean?), then $\ln(i) = i\frac{1}{2}\pi$ or $\pi = (2/i)\ln(i)$. The American mathematician Benjamin Peirce (1809–80) called this formal result a "mysterious formula." What an understatement! If, on the other hand, we raise both sides to the ith power (whatever *that* might mean), then we have

$$i^i = (e^{i\pi/2})^i = (\sqrt{-1})^{(\sqrt{-1})} = e^{i^2\pi/2} = e^{-\pi/2} = 0.2078 \ldots.$$

An imaginary number to an imaginary power can be *real*! Who could even have made up such an astonishing conclusion? As you will see in chapter 6 this isn't the end of the story, either—in fact, i^i has infinitely many real values, of which $e^{-\pi/2}$ is only one. Peirce's reference to the "mysterious formula," which he actually wrote as $\sqrt{-1}^{-\sqrt{-1}} = e^{\pi/2}$, appears in his 1866 book *Linear Associative Algebra*. An often told story has it that, after demonstrating the mysterious formula to one of his classes at Harvard, he declared "Gentlemen, this is surely true, it is absolutely paradoxical; we cannot understand it, and we don't know what it means. But we have proved it, and therefore we know it must be the truth." These words are certainly in keeping with the famous opening sentence of *Linear Associative Algebra*: "Mathematics is the science which draws necessary conclusions."

Expressions involving the products of trigonometric functions, such as $\sin(\alpha)\cos(\beta)$, often occur in scientific analyses,[5] and we can use the exponential connection to complex numbers to derive useful identities for such expressions. All of these derivations depend on noticing that, since $e^{i\theta} = \cos(\theta) + i\sin(\theta)$, we can then write $e^{-i\theta} = \cos(-\theta) + i\sin(-\theta) = \cos(\theta) - i\sin(\theta)$. If we add and subtract these two expressions for $e^{i\theta}$ and $e^{-i\theta}$ we can solve for $\cos(\theta)$ and $\sin(\theta)$ in terms of the exponentials:

$$\cos(\theta) = \frac{e^{i\theta} + e^{-i\theta}}{2} \text{ and } \sin(\theta) = \frac{e^{i\theta} - e^{-i\theta}}{2i}.$$

Now, to see how useful these expressions are, consider the product $\sin(\alpha)\cos(\beta)$, which we can write as

$$
\begin{aligned}
\sin(\alpha)\cos(\beta) &= \frac{e^{i\alpha} - e^{-i\alpha}}{2i} \cdot \frac{e^{i\beta} + e^{-i\beta}}{2} \\
&= \frac{e^{(\alpha+\beta)} - e^{-i(\alpha+\beta)} + e^{i(\alpha-\beta)} - e^{-i(\alpha-\beta)}}{4i} \\
&= \frac{2i\sin(\alpha+\beta) + 2i\sin(\alpha-\beta)}{4i} \\
&= \frac{1}{2}\sin(\alpha+\beta) + \frac{1}{2}\sin(\alpha-\beta).
\end{aligned}
$$

For the special case of $\alpha = \beta$ this identity reduces to

$$\sin(\alpha)\cos(\alpha) = \frac{1}{2}\sin(2\alpha),$$

a result derived earlier in this chapter in another way, i.e., using Wessel's idea of multiplying line segments. This same approach can be used to find identities for the other possible product expressions, $\sin(\alpha)\sin(\beta)$ and $\cos(\alpha)\cos(\beta)$.

Certain common trigonometric integrals yield easily when attacked with complex exponentials. Consider, for example, the problem of evaluating

$$\int_0^\pi \cos^n(\theta)\cos(n\theta)\, d\theta$$

where n is any non-negative integer. From the identity

$$2\cos(\theta)e^{i\theta} = 1 + e^{i2\theta}$$

it follows that

$$2^n\cos^n(\theta)e^{in\theta} = (1 + e^{i2\theta})^n.$$

Since this is an *identity* in θ, it remains an identity if we replace θ everywhere with $-\theta$. Remembering that the cosine is an even function, i.e., that $\cos(-\theta) = \cos(\theta)$, we then have

$$2^n\cos^n(\theta)e^{-in\theta} = (1 + e^{-i2\theta})^n.$$

Adding these two identities, then, gives us

$$2^n\cos^n(\theta)[e^{in\theta} + e^{-in\theta}] = 2^{n+1}\cos^n(\theta)\cos(n\theta)$$
$$= (1 + e^{i2\theta})^n + (1 + e^{-i2\theta})^n.$$

The left-hand side of the last equality looks just like the integrand of our integral, but what about the stuff on the right? It should be clear to you (I hope!) that, given enough time, we could multiply out each of the terms to get the two expansions

$$(1 + e^{i2\theta})^n = 1 + a_1e^{i2\theta} + a_2e^{i4\theta} + \cdots + a_ne^{in(2\theta)},$$
$$(1 + e^{-i2\theta})^n = 1 + b_1e^{-i2\theta} + b_2e^{-i4\theta} + \cdots + b_ne^{-in(2\theta)},$$

where the a's and the b's are numerical coefficients. Indeed, we could more easily calculate these coefficients by using the binomial theorem, but why bother? When we integrate both sides, i.e., when we write

$$\int_0^\pi 2^{n+1}\cos^n(\theta)\cos(n\theta)d\theta = \int_0^\pi \{\text{sum of the two expansions}\}\, d\theta,$$

then all of the integrals on the right of the form

$$\int_0^\pi e^{ik2\theta} d\theta, \; k = \pm 1, \pm 2, \ldots, \pm n$$

are zero. This important result is easy to prove formally by simply doing the easy exponential integral and plugging in the integration limits. Just remember that $e^{ik2\pi} = 1$ when k is any integer. Only the two leading 1's in the expansions result in nonzero integrals, then, to give

$$2^{n+1} \int_0^\pi \cos^n(\theta) \cos(n\theta) \, d\theta = \int_0^\pi 2 d\theta = 2\pi.$$

Thus, we have the wonderfully pretty result, for very little work, that

$$\int_0^\pi \cos^n(\theta) \cos(n\theta) \, d\theta = \frac{\pi}{2^n}, \; n = 0, 1, 2, \ldots.$$

As a little test for you on the above ideas, see if you can apply them to show that

$$\int_0^\pi \sin^{2n}(\theta) \, d\theta = \pi \frac{(2n)!}{2^{2n}(n!)^2}, \; n = 0, 1, 2, \ldots,$$

a famous result known by other means to John Wallis. Such calculations illustrate a famous saying attributed to the French mathematician Jacques Hadamard (1865–1963): "The shortest path between two truths in the real domain passes through the complex domain."

I'll conclude this section with a calculation that demonstrates the power of much of what has been discussed so far in this chapter. In addition, it will set the stage for the analysis I will do in chapter 7 when deriving an integral that plays a fundamental role in mathematics. To start, I want you to consider the general cyclotomic equation of even power, i.e., for n any positive integer, consider $z^{2n} - 1 = 0$. This equation has $2n$ roots evenly spaced in angle around the unit circle, with a separation of $2\pi/2n = \pi/n$ radians. Two of the roots are obvious, the real $z = \pm 1$; I think it equally obvious that there can be no other real roots. Therefore, the other $2n - 2 = 2(n - 1)$ roots must be complex. Half of these complex roots are on the upper part of the circle (they have positive imaginary part) and the other half (which are the conjugates of the first half) are on the lower part of the circle. All of the complex roots are compactly written (with all angles in radians) as $1 \angle (\pi \pm k\pi/n)$, $k = 1, 2, \ldots, n - 1$. Notice that $k = 0$ gives the real root $1 \angle \pi = -1$, while $k = n$

gives the other real root $1 \angle 0 = 1 \angle 2\pi = +1$. From all this, then, it follows that it must be possible to factor the cyclotomic equation as

$$z^{2n} - 1 = (z-1)(z+1)\prod_{k=1}^{n-1}\left\{z - 1 \angle \left(\pi + k\frac{\pi}{n}\right)\right\}\left\{z - 1 \angle \left(\pi - k\frac{\pi}{n}\right)\right\}$$

$$= (z-1)(z+1)\prod_{k=1}^{n-1}\left\{z^2 - z\left[1 \angle \left(\pi + k\frac{\pi}{n}\right) + 1 \angle \left(\pi - k\frac{\pi}{n}\right)\right] + 1 \angle 2\pi\right\}.$$

Now, $1 \angle 2\pi = 1$ and also

$$1 \angle \left(\pi + k\frac{\pi}{n}\right) + 1 \angle \left(\pi - k\frac{\pi}{n}\right) = e^{i(\pi + k\pi/n)} + e^{i(\pi - k\pi/n)}$$

$$= e^{i\pi}(e^{ik\pi/n} + e^{-ik\pi/n})$$

$$= -2\cos\left(k\frac{\pi}{n}\right).$$

Thus,

$$z^{2n} - 1 = (z-1)(z+1)\prod_{k=1}^{n-1}\left\{z^2 + 2z\cos\left(k\frac{\pi}{n}\right) + 1\right\}$$

or, factoring a z out of each of the $n - 1$ terms in the braces on the right,

$$z^{2n} - 1 = (z^2 - 1)z^{n-1}\prod_{k=1}^{n-1}\left\{z + 2\cos\left(k\frac{\pi}{n}\right) + z^{-1}\right\}.$$

Dividing through by z^n gives

$$z^n - z^{-n} = (z - z^{-1})\prod_{k=}^{n-1}\left\{z + 2\cos\left(k\frac{\pi}{n}\right) + z^{-1}\right\}.$$

If we next set the complex variable z to be any point on the unit circle, then we can write $z = \cos(\theta) + i\sin(\theta)$. Recall from section 3.1 that $z^{-1} = \cos(\theta) - i\sin(\theta)$, and so from De Moivre's theorem $z^n = \cos(n\theta) + i\sin(n\theta)$ as well as $z^{-n} = \cos(n\theta) - i\sin(n\theta)$. Thus,

$$2i\sin(n\theta) = 2i\sin(\theta)\prod_{k=1}^{n-1}\left\{2\cos(\theta) + 2\cos\left(k\frac{\pi}{n}\right)\right\},$$

$$\frac{\sin(n\theta)}{\sin(\theta)} = 2^{n-1}\prod_{k=1}^{n-1}\left\{\cos(\theta) + \cos\left(k\frac{\pi}{n}\right)\right\}.$$

If we then let $\theta \to 0$ we have $\lim_{\theta \to 0}\sin(n\theta)/\sin(\theta) = n$, and of course $\lim_{\theta \to 0}\cos(\theta) = 1$. Again, as I promised in section 3.2 during the derivation of

71

Viète's infinite product formula for π, the first of these two statements will be formally derived in chapter 6. So we arrive at

$$n = 2^{n-1} \prod_{k=1}^{n-1} \left\{ 1 + \cos\left(k\frac{\pi}{n} \right) \right\}.$$

Recall now the half-angle formula from section 3.2, i.e., for α any angle

$$\cos\left(\frac{1}{2}\alpha \right) = \sqrt{\frac{1 + \cos(\alpha)}{2}},$$

which says that $\cos(\alpha) = 2\cos^2(\frac{1}{2}\alpha) - 1$. Thus,

$$1 + \cos\left(k\frac{\pi}{n} \right) = 2\cos^2\left(\frac{k\pi}{2n} \right)$$

and so

$$n = 2^{n-1} \prod_{k=1}^{n-1} 2\cos^2\left(\frac{k\pi}{2n} \right) = 2^{2(n-1)} \prod_{k=1}^{n-1} \cos^2\left(\frac{k\pi}{2n} \right).$$

Taking the square root of both sides—which we can do without having any sign ambiguity concerns because $\cos(k\pi/2n) > 0$ for all k from 1 to $n - 1$—we finally arrive at

$$\prod_{k=1}^{n-1} \cos\left(\frac{k\pi}{2n} \right) = \frac{\sqrt{n}}{2^{n-1}}.$$

This is certainly a pretty formula—I think the $\sqrt{n}$ in the numerator is totally unexpected—but is it correct? Well, we can always check the formula for any specific value of n. Let us try $n = 5$. Then the formula claims that

$$\prod_{k=1}^{4} \cos\left(\frac{k\pi}{10} \right) = \frac{1}{16}\sqrt{5} = 0.1397543,$$

and in fact if you run the product of the four cosine factors on the left through your calculator you'll find that is just what you will indeed get. And, by the way, if you simply define a new index variable on the product as $j = n - k$ (so that as k runs from 1 to $n - 1$ then j runs from $n - 1$ to 1), and if you recall that $\sin(\theta) = \cos(\frac{1}{2}\pi - \theta)$, then it is easy to see that we also have the result

$$\prod_{k=1}^{n-1} \sin\left(\frac{k\pi}{2n} \right) = \frac{\sqrt{n}}{2^{n-1}}.$$

There are no $\sqrt{-1}$ terms left in the final formulas, but can you even begin to imagine how to derive them *without* $\sqrt{-1}$? I can't.

3.4 ARGAND

As I stated before, brilliant as Wessel's contribution was, it simply was not read—and would not be—until unearthed from the musty literature a century after the fact. Wessel's ideas were, however, "in the air" and within a decade of the original presentation of his paper to the Danish Academy they were rediscovered. Indeed, the year 1806 actually saw two publications that, more or less, put forth Wessel's complex plane and his association of the vertical axis with the axis of imaginaries.

The first of these two writers, the Swiss-born Jean-Robert Argand (1768–1822), was from as unlikely a background as was Wessel. Essentially nothing is known of Argand's early life, but he probably was not formally trained in mathematics—in 1806, when nearly forty, he was laboring in obscurity as a Parisian bookkeeper. In 1876 the French mathematician Guillaume-Jules Hoüel (1823–86) reprinted Argand's pamphlet, and in his introductory preface tells of his attempt to track down details of Argand's life. It was through Hoüel's efforts that Argand's registry of birth was found. He found very little more beyond that, however. Hoüel ended his brief account with the following poignant words: "If we add to this that, about 1813, Argand lived at Paris *rue de Gentilly, No. 12,* as indicated in his own handwriting on the cover of [a copy of his pamphlet] we shall have stated all we have been able to learn of this original man, whose modest life will remain unknown, but whose services to science Hamilton and Cauchy have deemed worthy the gratitude of posterity."

Despite his humble origin, in 1806 Argand had his work on complex numbers—in which the idea of the modulus is first introduced—published in a small print run by a private press. Since he probably meant to give copies away free to friends and correspondents, who would of course know who the author was, he did not even put his name on the title page. This work, titled *Essay on the Geometrical Interpretation of Imaginary Quantities,* was almost surely doomed to disappear even faster than had Wessel's, except for subsequent events of an astonishing nature.

One who received a copy of Argand's work was the great French mathematician Adrien-Marie Legendre (1752–1833), who in turn mentioned it in a letter to Francois Francais (1768–1810), a professor of mathematics whose military background led him naturally into mathematical problems related to artillery; for example, the use of calculus to study the motion of projectiles through air. When he died, his younger brother Jacques (1775–1833) inherited the older brother's papers. Jacques, too, had an extensive military background and, from 1811 until his death, he was professor of military art at the Ecole Impériale d'Application du Génie et de l'Artillerie in Metz. Also, like his older brother, Jacques was a mathematician and, while going through his late brother's papers, he found Legendre's 1806 letter describing the mathematics in Argand's pamphlet—but he didn't know Argand's name as Legendre had failed to mention it in the letter.

Stimulated by the ideas he read about in that letter, Jacques published an article in an 1813 issue of the journal *Annales de Mathématiques* giving the basics of the geometry of complex numbers. In the last paragraph of his paper, however, Francais acknowledged his debt to Legendre's letter and, further, urged the anonymous author whose work Legendre discussed in that letter to come forward. Happily, Argand learned of this plea and his reply appeared in the very next issue of the journal. Accompanying Argand's reply was a short note by Francais, in which he declared Argand to be the first to have developed the geometry of imaginaries and that he was pleased to so acknowledge this priority (neither man had, of course, heard of Wessel). What a happy story this is, particularly when compared to the sorry mess of tangled claims of priority that still surrounds the Tartaglia/Cardan controversy over the cubic.

With the publication of Argand's ideas in a recognized journal of mathematics, some controversy erupted between Argand and Francais on one side and the French mathematician Francois-Joseph Servois (1767–1847) on the other. A man with a military background who, like Francais, taught for a while at the artillery school in Metz, Servois felt that a geometric interpretation of algebraic ideas somehow made them impure. For example, in a letter to the *Annales* he wrote "I confess that I do not yet see in this notation anything but a geometric mask applied to analytic forms the direct use of which seems to me simple and more expeditious."[6] Curiously, this debate remained more or less courteous and probably, if anything, helped to attract some attention to Argand's ideas. Even more curiously, Wessel (who was still alive) heard nothing of these doings in France, and nobody else remembered, if they ever knew, that Wessel had done it all twenty years before. Wessel and Argand went to their graves only four years apart, with neither having learned of the other and with most of the world ignorant of both.

In Hoüel's 1876 reprint of Argand's work the following ironic testimonial was included in the introductory preface, words quoted from the recently deceased young German mathematician Hermann Hankel (1839–73): "The first to show how to represent the imaginary form $A + Bi$ by points in a plane, and to give rules for their geometric addition and multiplication, was Argand. . . . unless some older work is discovered, Argand must be regarded as the true founder of the theory of complex quantities in a plane." Two decades later, of course, Wessel's "older work" *was* discovered.

3.5 BUÉE

The second rediscovery in 1806 of Wessel's ideas came from an even more obscure writer than Argand, the French Abbé Adrien-Quentin Buée (1748–1826), who published, in French, a very long paper in the *Philosophical Transactions* of the Royal Society of London. Unlike the workmanlike style of both Wessel and Argand, Buée's work has a decidedly mystical air to it, and it must have seemed quite strange to all who managed to get through its sixty-five pages. His geometric understanding of $\sqrt{-1}$ is clear when he writes "$\sqrt{-1}$ is the sign of perpendicularity." Matters become somewhat suspect, however, when after failing to define multiplication of line segments (Wessel's central contribution) Buée then states that if t represents *future* time and if $-t$ represents *past* time, then the *present* time is composed of $t \times \frac{1}{2}\sqrt{-1}$ and $-t \times \frac{1}{2}\sqrt{-1}$.

As one Harvard mathematician wrote on this a century later, "It is vain to speculate after this lapse of time as to why such a memoir was accepted by the Society. Was the good Abbé an *émigré* whom the British delighted to honor the year that they defeated his non-emigrated countrymen at Trafalgar? Some such reason there must have been, as the intrinsic worth of the memoir would never recommend it for publication."[7] Five years later an English mathematician offered a somewhat less snide assessment when he wrote "It must be stated that this paper serves to indicate little more than the fact that the Abbé's ideas concerning imaginaries are decidedly unorthodox."[8] These two comments are generally correct, but it is also true that the Irish mathematical genius William Rowan Hamilton seems, in his correspondence, to have had a favorable, or at least a tolerant, opinion of Buée's paper.

In fact, while Buée's paper does have a slightly mystical flavor to it, he did not write it simply to proselytize his own peculiar ideas. Rather, he wrote in an attempt to respond to questions raised in a book written just three years before, by the respected French mathematician and statesman Lazare Carnot (1753–1823). (Carnot—whose son, Sadi, is known today as the "father of modern thermodynamics"—was greatly troubled by the mysteries of $\sqrt{-1}$ and, in his 1803 book *Géométrie de Position,* he formulated many interesting questions, one of which is discussed in box 3.1.) Buée's paper is speculative, yes, indeed it is even "strange," but it was certainly written in a scholarly search for truth. I don't think Buée was, in any sense of the word, a quack, and certainly not deserving of ridicule.

Indeed, Buée's paper struck at least one well-known and respected French mathematician at the time as sufficiently meritorious that he[9] suggested Argand's work could have been written after Argand had seen Buée's paper.

BOX 3.1

CARNOT'S PROBLEM OF DIVIDING A LINE SEGMENT

In his 1803 book *Géométrie de Position* Carnot asks the following question: given a line segment *AB,* of length *a,* how can it be divided into two shorter segments so that the product of their lengths is equal to one-half the square of the original length?* Carnot solves this by letting the two new lengths be denoted by *x* and *a* − *x.* Then,

$$x(a-x) = \frac{1}{2}a^2$$

and this is easily solved to give

$$x = \frac{1}{2}a \pm i\frac{1}{2}a.$$

Just as discussed back in section 2.1, the appearance of a complex so-lution was interpreted by Carnot to mean that the point on the line segment defined by *x* (the distance from *A* to the point) does not lie between *A* and *B.* That is, it is not physically possible to divide *AB* as required by the given constraint.

Buée, on the other hand, while agreeing that the dividing point could not be between *A* and *B,* imagined that the imaginary part of *x* meant the point is located off to the side of *AB,* i.e., at a distance of ½*a* from the midpoint (because the real part of *x* is ½*a*) of *AB.* Hence Buée's interpretation of $i = \sqrt{-1}$ as meaning perpendicularity. Buée offers nothing to support this idea, however, unlike Wessel's analysis, in which the same conclusion follows logically.

* I have been unable to obtain a copy of Carnot's book, and so the discussion in box 3.1 is based on the tutorial essay by Alexander MacFarlane, "On the Imaginary of Alge-bra," *Proceedings of the American Association for the Advancement of Science* 41 (1892):33–55.

This Argand hotly rejected, arguing that there is generally a significant lapse between the date a journal bears and when it actually appears, and since his pamphlet was published in 1806 then he could not have seen Buée's work prior to writing his own. As Argand put it, obviously insulted at such a com-ment hinting transparently at plagiarism, "This is sufficient to prove that if the contribution of Buée was wholly his own, as is quite possible, then it is also quite certain that I could have had no knowledge of his paper when my treatise appeared." In any case, the two works are significantly different in presentation, with Argand's the clear superior.

But this is not to say that Argand, too, didn't have his problems with complex numbers. For example, he was at first convinced that $(\sqrt{-1})^{\sqrt{-1}}$ could not be written in the two-dimensional form $a + ib$, but rather would require a three-dimensional space. In one of the exchanges in the *Annales,* Francais pointed out that Argand was wrong; or as he diplomatically put it, Argand's claim was "simply a conjecture open to serious objection." In fact, he went on to derive the even more general formula

$$(c\sqrt{-1})^{d\sqrt{-1}} = e^{-d\pi/2}\{\cos[d\ \ln(c)] + \sqrt{-1}\ \sin[d\ \ln(c)]\}.$$

For $c = d = 1$, this reduces to $i^i = e^{-\pi/2}$ which is obviously purely real, and so $(\sqrt{-1})^{\sqrt{-1}}$ needs only the one-dimensional real axis to find its home— not even the full two dimensions of the complex plane are needed and certainly not the three of space. Box 3.2 shows how Francais' generalization can be made even *more* general. And box 3.3 should keep you sleepless for a few nights!

3.6 REDISCOVERY REDUX

After 1814 Argand's work drifted toward the same oblivion as had Wessel's, and matters had to be rediscovered yet again. Just as 1806 had been a year of simultaneous discovery for Argand and Buée, 1828 also saw the publication of two books, one in Cambridge, England and the other in Paris. The latter, by C. V. Mourey, I have been unable to obtain,[10] but the other, by the Reverend John Warren (1796–1852), a Fellow of Jesus College of Cambridge University, was a solid, formula-packed presentation of complex numbers (*A Treatise on the Geometrical Representation of the Square Roots of Negative Quantities*). Warren's book would perhaps have its greatest impact not by what it had to say, but by how it said it, and in how its geometrical approach was rejected by William Rowan Hamilton. Hamilton was inspired by Warren's book to think he could do better, and as an already famous man his work, so one might think, would *not* be forgotten.

William Rowan Hamilton (1805–65) was born in Dublin, Ireland, and by the time he died not far from where he began his life he had left his mark all over mathematical physics. He got off to a rather odd start, however. A youth of almost unbelievably misguided precocity, he devoted many of his early years to gaining proficiency in multiple languages—Latin, Greek, and

BOX 3.2

RAISING A COMPLEX NUMBER TO A COMPLEX POWER

Francais' rebuttal to Argand's claim that i^i cannot itself be written as a complex number could have been even more general than it was. He could have calculated

$$(a+b\sqrt{-1})^{c+d\sqrt{-1}}$$

and shown that even this more "complex" expression can be written as the sum of a real part and an imaginary part. Euler did this in 1749, when Wessel was four years old (and Argand was -19) and the French mathematician and philosopher Jean le Rond d'Alembert (1717–83) claimed to have done it even earlier. Argand's i^i would then have simply been the special case of $a = c = 0$ and $b = d = 1$. Can *you* do this? As a hint, I will calculate $(1 + i)^{1+i}$ and that should provide you with enough clues for the general approach. To start, write $1 + i = \sqrt{2} \angle (\pi/4 + 2\pi k)$, where k is any integer. Then, using the identity $x = e^{\ln(x)}$,

$$(1+i)^{1+i} = e^{\ln(1+i)^{1+i}} = e^{(1+i)\ln(1+i)} = e^{(1+i)\ln\{\sqrt{2}\angle(\pi/4+2\pi k)\}}$$

$$= e^{(1+i)\ln\{\sqrt{2}e^{i(\pi/4+2\pi k)}\}} = e^{(1+i)\{\ln(\sqrt{2})+i\pi(1/4+2k)\}}$$

$$= e^{\{\ln(\sqrt{2})-\pi(1/4+2k)\}}e^{i\{\ln(\sqrt{2})+\pi(1/4+2k)\}}.$$

The so-called principal value for $(1 + i)^{1+i}$ is the $k = 0$ case, which gives

$$e^{\{\ln(\sqrt{2})-\pi/4\}}e^{i\{\ln(\sqrt{2})+\pi/4\}} = e^{-\pi/4}\sqrt{2}\left[\cos\left\{\ln(\sqrt{2})+\frac{\pi}{4}\right\}+i\sin\left\{\ln(\sqrt{2})+\frac{\pi}{4}\right\}\right]$$

$$= 0.2739 \ldots + i0.5837 \ldots.$$

Using the same approach, see if you can show that $1^i = e^{-2\pi k}$. Thus, while the principal value of $1^i = 1$, the common saying of "one to any power is always one" is not true in general.

Hebrew by age five, to which by age nine he had added Persian, Arabic, Sanskrit, etc., etc. As the historian of mathematics E. T. Bell wrote of all this effort, "Good God! What was the sense of it all?"[11] By the end of 1824, however, discipline appeared and mathematics had become his passion to the point that he presented a paper on optics to the Royal Irish Academy. The

BOX 3.3

EXPONENTIAL MADNESS

If you think exponentials are pretty straightforward at this point, then consider the following infuriating little puzzle. It was proposed in 1827 by a one-time cattle herder, the self-taught Danish mathematician Thomas Clausen (1801–85). Gauss himself declared Clausen to have "outstanding talents" and, as this puzzle shows, perhaps even diabolic ones. Clausen wrote the puzzle as a challenge problem, and variations of it periodically appear in the problem sections of mathematics journals to this day.

$$e^{i2\pi n} = 1 \text{ for } n = 0, \pm 1, \pm 2, \ldots$$

$$e \cdot e^{i2\pi n} = e = e^{1+i2\pi n}$$

$$e^{1+i2\pi n} = \{e^{1+i2\pi n}\}^{(1+i2\pi n)}$$

$$e = e^{(1+i2\pi n)^2} = e^{1+i4\pi n - 4\pi^2 n^2} = e^{1+i4\pi n}e^{-4\pi^2 n^2}$$

$$e^{1+i4\pi n} = e$$

$$e = ee^{-4\pi^2 n^2}$$

$$1 = e^{-4\pi^2 n^2}$$

The last statement is true only for $n = 0$. And yet, we started with a statement true for all integer n and then performed operations that appear to always be valid. What happened?

Academy did have some trouble understanding everything that the still teen-age Hamilton had to say, but by 1836 he had learned how to express himself well enough that the Royal Society awarded him its Royal Medal that year for some important optical studies.

Still, while his name lives in physics today in a quantity called the Hamiltonian, a concept studied by every undergraduate physics student learning about the science of dynamics, little is remembered of the man. His early work in optics and wave motion, which led to the formulation of the Hamiltonian, has a particular irony to it, too, when one realizes that a century later the emerging quantum-wave mechanics found Hamilton's ideas fundamentally important. Ironic, because it is the names of others, Louis de Broglie and Erwin Schrödinger, in particular, that are associated with wave mechanics, not Hamilton's.

In mathematics, too, Hamilton made a big splash during his lifetime only to have his ghost watch his association with it all fade to a mere shadow of the original glory. Hamilton's work in complex numbers is the prime example of this fate. In 1829 a friend urged him to read Warren's book, published the year before. He did and Hamilton, who felt that algebra—with its $\sqrt{-1}$—was separate and distinct from geometry, rejected that book's geometric interpretation of complex numbers. Rather, he felt that $\sqrt{-1}$ should have a purely algebraic interpretation. As he wrote much later, in 1853, "I . . . felt dissatisfied with any view which should not give to [imaginaries] from the outset a clear interpretation and *meaning*; and wished that this should be done, for square roots of negatives, without introducing considerations so *expressly geometrical* as those which involved the conception of an angle."[12]

Hamilton felt that, just as geometry is the science of space and just as that science had found its mathematical expression in Euclid's *Elements,* algebra too should be the science of something in our physical existence. He decided the "something" must be time, an idea he picked up from Kant's philosophy, and declared that time must be what algebra—with its $\sqrt{-1}$—is all about. In associating time with $\sqrt{-1}$ Hamilton was, of course, not the first; recall Buée, whose 1806 paper Hamilton had read. Hamilton would certainly have been less sure of himself on this line of reasoning if he had known that there is no *one* science of geometry, i.e., Euclid's geometry is only one of many possible, self-consistent geometries. Euclidean geometry does seem to describe our local surroundings quite well, but on a cosmic scale physicists have turned to non-Euclidean geometries to describe what nearly everybody who has seen *Star Trek* has heard of these days—the geometry of *curved spacetime*. So, just as there is no one science of space, there is no one science of algebra either— another algebra that is commonly used today is Boolean algebra, the logical algebra of sets used to design digital machines, like computers—and they can't *all* be about time!

Hamilton knew nothing of such matters, however, and so in June of 1835 he presented to the Irish Academy a paper titled "Theory of Conjugate Functions or Algebraic Couples: with a Preliminary Essay on Algebra as a Science of Pure Time." I will skip over the metaphysics of algebra as the science of time, and simply tell you about the mathematics. Hamilton defined *ordered pairs* of real numbers, written as (a,b), to be a *couple*. He defined the addition and multiplication of two couples as follows:

$$(a,b) + (c,d) = (a + c, \, b + d),$$
$$(a,b)(c,d) = (ac - bd, \, bc + ad).$$

Note very carefully that these are *definitions,* and as such require no further explanation. In fact, however, it is painfully clear that Hamilton was motivated to make these particular definitions because he already knew that that is how complex numbers "work." That is, his mathematically "pure and abstract" couple (a,b) is just another way to write $a + ib$. Hamilton, however, felt his notation somehow better since it avoids the use of the "absurd" $\sqrt{-1}$. Still, as a wit once put it, "A rose is a rose, and a pig by any other name is still a pig." There is an amusing comment by Hamilton himself on the fact that his abstract definitions are really not arbitrary at all: "Persons who have read with attention the foregoing remarks of this theory . . . will see that these definitions are really not arbitrarily chosen." Indeed not.

In any case, continuing with the analogy to complex numbers, Hamilton wrote the purely real number a as the couple $(a,0)$. This is, of course, exactly what you do when you write a as the abbreviated form of $a + 0\sqrt{-1}$. So, formally, we have from the multiplication definition that

$$a(c,d) = (a,0)(c,d) = (ac,ad)$$

and, in particular, if $a = -1$ then

$$-1(c,d) = -(c,d) = (-1,0)(c,d) = (-c,-d).$$

Now,

$$(0,1)^2(c,d) = (0,1)(0,1)(c,d) = (0,1)(-d,c)$$
$$= (-c,-d) = -(c,d).$$

Thus, $(0,1)^2 = -1$ or, just as one would expect, we have $(0,1) = \sqrt{-1}$. There should be no surprise here that $(0,1) = i = \sqrt{-1}$, because we started with $(a,b) = a + ib$.

There is not just a little irony in Hamilton's aversion to a geometrical view of complex numbers. I say this because Hamilton of course knew of the rotational property of $\sqrt{-1}$. Indeed, it was that *geometrical* knowledge that led him on to his next mathematical quest, one that was to obsess him for the rest of his life. Just as $\sqrt{-1}$ rotates vectors in the complex plane, Hamilton wondered what would rotate vectors in three-dimensional space. This led him to his discovery of quaternions or *hypercomplex* numbers, a story I will not tell here.[13]

3.7 GAUSS

By the time Hamilton published his work on couples the geometric interpretation had already been given the stamp of approval by the intellectual giant Carl Friedrich Gauss (1777–1855). In April of 1831, four years before Hamilton, Gauss had already presented his own geometric ideas on complex numbers in a memoir to the Royal Society of Göttingen. In fact, Gauss had been in possession of these concepts in 1796 (before Wessel) and had used them to reproduce, without Gauss' knowledge, Wessel's results. But, like so many other of his works of genius, he had neglected to publish until he felt he had everything "just right." In an 1812 letter to the French mathematician Pierre Laplace, for example, Gauss wrote "I have in my papers many things for which I could perhaps lose the priority of publication, but you know, I prefer to let things ripen."[14]

In 1831 he had finally reached a sufficiently ripened state concerning complex numbers (the term *complex* is his), and it was Gauss' enormous reputation that at last carried the day. Gauss' understanding of complex numbers did evolve over time. For example, at the time of the writing of his dissertation he believed there might be an endless hierarchy of complex numbers, i.e., that the complex numbers might not be complete. He called these ever more complex numbers "vera umbrae umbra," or "veritable shadows of shadows." Later, of course, he realized that this is not so. In his honor the complex plane is sometimes referred to as the "Gaussian plane," except in France where one is equally likely to read of the "Argand plane." Complex numbers of the form $a + ib$ with a and b both integers are called Gaussian integers. After Gauss, $\sqrt{-1}$ was accepted as a legitimate symbol and, on the occasion of the fifty-year jubilee (in 1849) celebrating his doctorate, the congratulatory address told him "You have made possible the impossible."

As Gauss wrote in 1831, "If this subject has hitherto been considered from the wrong viewpoint and thus enveloped in mystery and surrounded by darkness, it is largely an unsuitable terminology which should be blamed. Had $+1$, -1 and $\sqrt{-1}$, instead of being called positive, negative and imaginary (or worse still impossible) unity, been given the names, say, of direct, inverse and lateral unity, there would hardly have been any scope for such obscurity." It is ironic that in that same year Hamilton's friend Augustus De Morgan wrote, in his book *On the Study and Difficulties of Mathematics,* "We have shown the symbol $\sqrt{-1}$ to be void of meaning, or rather self-contradictory and absurd." Years later De Morgan was still a skeptic; in his 1849 book *Trigonometry and Double Algebra* he wrote "The student, if he should hereafter inquire into the assertions of different writers, who contend for what each of them considers as *the* explanation of $\sqrt{-1}$, will do well to substitute the indefinite article." De Morgan was too much the skeptic, but he was not ignorant—in that same book he managed to overcome his distaste for $\sqrt{-1}$ and perform essentially the general calculations given in box 3.2. As one modern writer has put it, "At the beginning of their history, complex numbers $a + b\sqrt{-1}$ were considered to be 'impossible numbers,' tolerated only in a limited algebraic domain because they seemed useful in the solution of cubic equations. *But their significance turned out to be geometric* [my emphasis] and ultimately led to the unification of algebraic functions with conformal mapping, potential theory, and another 'impossible' field, noneuclidean geometry. This [geometric] resolution of the paradox of $\sqrt{-1}$ was so powerful, unexpected, and beautiful that only the word 'miracle' seems adequate to describe it."[15]

There *were*, I should tell you, some who still resisted. While writing in the Cambridge Philosophical Society's *Transactions,* for example, the English mathematician George Airy (1801–92), who was the Astronomer Royal from 1835 to 1881, declared "I have not the smallest confidence in any result which is essentially obtained by the use of imaginary symbols." Five years after Gauss' jubilee, the English logician George Boole (1815–64) called $\sqrt{-1}$ an "uninterpretable symbol" in his 1854 masterpiece *An Investigation of the Laws of Thought.* And finally, as late as the 1880s, the situation was such that one of the top mathematics students in England would recall, decades later, that "it was an age when the use of $\sqrt{-1}$ was suspect at Cambridge even in trigonometrical formulae. . . . The imaginary *i* was suspiciously regarded as an untrustworthy intruder."[16] Change comes hard, even for mathematicians.

Using Complex Numbers

4.1 COMPLEX NUMBERS AS VECTORS

In this chapter, and in the next, I will show you some specific examples or case studies of the application of complex numbers to the solution of interesting problems in mathematics and applied science. Most of the underlying theory in this chapter will be based on the elementary idea that complex numbers can represent vectors, i.e., quantities with magnitude and direction, in the complex plane. Indeed, complex number arithmetic can be interpreted as sequences of vector manipulations.

The addition and subtraction of vectors is routinely taught in high school physics and, as shown in figure 4.1, the ideas are quite straightforward. In that figure the complex numbers $2 - i3$ and $3 + i4$ are shown as vectors, and their sum is calculated using the well-known parallelogram rule of adding "head-to-tail" to get $5 + i$. To subtract, say, $2 - i3$ from $3 + i4$, we first form $-(2 - i3) = -2 + i3$ (which, as a multiplication of $2 - i3$ by $-1 = 1 \angle 180°$ is simply a $180°$ clockwise rotation of $2 - i3$) and then add as before to get $1 + i7$. This is shown in figure 4.2. Notice that the $180°$ rotation is equivalent to extending $2 - i3$ backwards along its own direction, through the origin.

Less familiar than addition and subtraction, perhaps, but really no more difficult, is the geometrical vector multiplication and division of complex numbers.[1] To see how multiplication works, you need only recall that it is multiple addition. For example, for real numbers, $3 \times 5 = 3 + 3 + 3 + 3 + 3 = 5 + 5 + 5$. That is, we can either add 5 three times or add 3 five times. This elementary observation can be easily extended to complex number/vector multiplication, as well. So, suppose we want to calculate $(2 - i3)(3 + i2)$. To do this geometrically, arbitrarily pick either one of the factors, say $3 + i2$, and add it to itself $2 - i3$ times. This isn't as crazy as it perhaps sounds! Just notice that $(2 - i3)(3 + i2) = 2(3 + i2) - i3(3 + i2)$. This tells us to first draw the vector $3 + i2$, and then make it twice as long. Call the result $V1$, for "vector one." Then draw the vector $3 + i2$, make it three times as long, and, finally, rotate the result by $-90°$ because of the factor $-i = 1 \angle -90°$. Call the result of these operations $V2$. The answer we want is just $V1 + V2$, a vector addition that we already know how to do. The entire process is shown in figure 4.3, where the rotation operation on $3 + i2$ can be thought of as

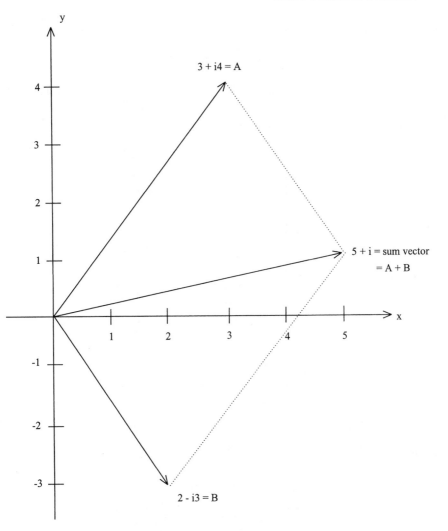

Figure 4.1. Vector addition.

defining a new set of coordinate axes, x' and y', in which the final addition operation is performed. In the same way, division of complex number vectors can be interpreted as multiple subtraction.

4.2 DOING GEOMETRY WITH COMPLEX VECTOR ALGEBRA

One elementary but most useful application of treating complex numbers as vectors is in proving geometric theorems. To illustrate this I will first demon-

85

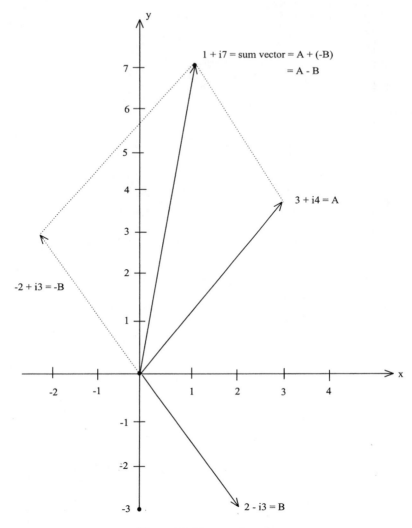

Figure 4.2. Vector subtraction.

strate the technique by proving a theorem from elementary Euclidean geometry that, using the traditional axiomatic approach, is not so easy to do. I think,
however, that you will find the complex vector proof to be almost transparent.
Then, with that done, I will show you a beautifully short complex vector proof
of a simple version of an elegant theorem due to Roger Cotes, about whom
much more will be said in chapter 6.

To begin, let me ask you to consider the following fundamental assertion
which I'll soon prove for you. Let me call it assertion 1, as eventually there

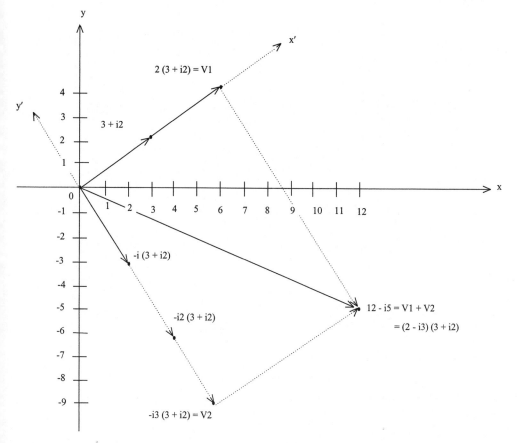

Figure 4.3. Vector multiplication.

will be an assertion 2. So, assertion 1: Suppose you are given two points in the complex plane, P_1 and P_2, with coordinates $z_1 = x_1 + iy_1$ and $z_2 = x_2 + iy_2$, respectively, as shown in figure 4.4. Further, let P be an arbitrary point on the line segment connecting P_1 and P_2. Then, if P divides the line segment P_1P_2 into two (obviously shorter) line segments P_1P and PP_2, where $P_1P/PP_2 = \lambda$, then we can write the location of P as

$$z = \frac{z_1 + \lambda z_2}{1 + \lambda}.$$

In particular, if $\lambda = 1$ then P is the midpoint of P_1P_2 and

$$z = \frac{z_1 + z_2}{2},$$

87

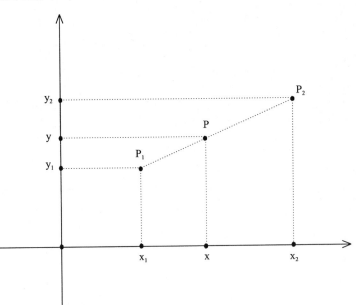

Figure 4.4. The geometry of assertion 1.

i.e., z is the average of z_1 and z_2, and P bisects P_1P_2. And if $\lambda = 2$, then P is a trisection point of P_1P_2 and

$$z = \frac{z_1 + 2z_2}{3}.$$

Notice that this trisection point is closer to P_2 than it is to P_1. The other trisection point, the one closer to P_1 than it is to P_2, occurs for $\lambda = \frac{1}{2}$ which gives

$$z = \frac{2z_1 + z_2}{3}.$$

Assertion 1 is actually quite easy to prove, using figure 4.4. Simply notice that since P_1 and P_2 are on the same line segment, and by the definition of λ, we can write

$$\frac{x - x_1}{y - y_1} = \frac{x_2 - x}{y_2 - y}, \quad \frac{x - x_1}{x_2 - x} = \lambda, \quad \frac{y - y_1}{y_2 - y} = \lambda.$$

These three expressions are easily manipulated to give

$$x = \frac{x_1 + \lambda x_2}{1 + \lambda} \text{ and } y = \frac{y_1 + \lambda y_2}{1 + \lambda},$$

which is assertion 1 with $z = x + iy$.

Now we will use assertion 1 to prove the following claim: the medians of any triangle meet in a point P which is two-thirds of the way from each vertex to the opposite side, as shown in figure 4.5. Here is the proof. Each median from a vertex terminates, by definition, at the midpoint of the side opposite the vertex. The locations of the terminating points of the three medians, from assertion 1 (using $\lambda = 1$ for bisecting points), are as shown in figure 4.5 where the locations of the three vertices A, B, and C are z_1, z_2, and z_3, respectively. Now, for each median line segment, we can compute the point that is two-thirds of the way from the appropriate vertex, again using assertion 1 with $\lambda = 2$. This gives the trisection point closer to the terminating point, which is what we want. Let us call these three points P_A, P_B, and P_C, and then show they are the same point as claimed by the theorem. Thus,

$$P_A = \frac{1}{3}\left[z_1 + 2 \cdot \frac{1}{2}(z_2 + z_3) \right] = \frac{1}{3}(z_1 + z_2 + z_3),$$

$$P_B = \frac{1}{3}\left[z_2 + 2 \cdot \frac{1}{2}(z_1 + z_3) \right] = \frac{1}{3}(z_2 + z_1 + z_3),$$

$$P_C = \frac{1}{3}\left[z_3 + 2 \cdot \frac{1}{2}(z_1 + z_2) \right] = \frac{1}{3}(z_3 + z_1 + z_2),$$

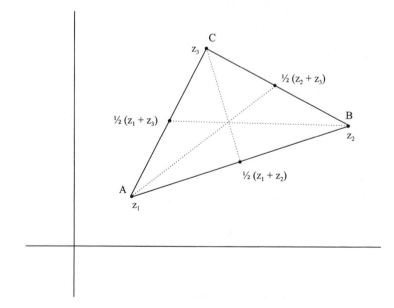

Figure 4.5. A geometric theorem.

and obviously $P_A = P_B = P_C$. So, we are done! Wasn't that easy? Try creating a traditional high school geometry proof and I think you will find it much more difficult.

Before turning to Cotes' theorem, here is assertion 2. The distance between two points z_1 and z_2 in the complex plane is $|z_1 - z_2| = \sqrt{(x_1 - x_2)^2 + (y_1 - y_2)^2}$. The proof of this is simply to remember the Pythagorean theorem. Now let us use this to discuss Cotes' theorem. Cotes' theorem: If a regular n-gon is inscribed in a circle with radius r, and if the point P lies on one of the radii from the center of the circle, i.e., of the n-gon, to one of the vertices at distance a from the center, then the product of the distances between P and all of the vertices is $r^n - a^n$ or $a^n - r^n$, depending on whether P is inside the circle ($a < r$) or outside the circle ($a > r$), respectively. Figure 4.6 shows the geometry involved for the case of $n = 4$ and for P inside the circle.

Cotes was led to this theorem because of a claim made by Leibniz in his 1702 book *Acta Eruditorum*. There he observed that the integrals

$$\int \frac{dx}{x + a} \text{ and } \int \frac{dx}{x^2 + a^2}$$

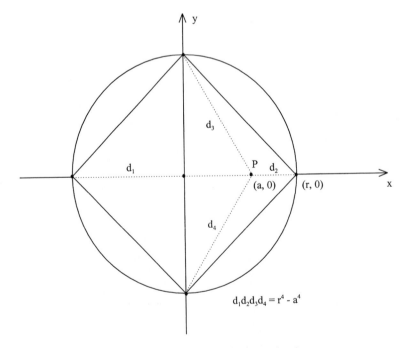

Figure 4.6. Cotes' theorem, for a regular 4-gon.

are expressible in terms of the logarithmic and trigonometric functions, respectively, but the next such integrals of

$$\int \frac{dx}{x^4 + a^4} \text{ and } \int \frac{dx}{x^8 + a^8}$$

cannot be so expressed. Cotes responded to this (incorrect) claim by studying how to factor $x^n \pm a^n$ and, indeed, how to construct the factors geometrically. Hence his theorem, which he only stated and did not prove. It was not until 1722 that a proof was finally published by Henry Pemberton (1694–1771), who was editor of the third edition of Newton's *Principia*. John Bernoulli described Pemberton's proof as "long, tedious and intricate," but that's because Pemberton didn't use complex numbers! We know of Cotes' theorem only because his cousin, the physicist Robert Smith (1689–1768), included it in *Harmonia Mensurarum* after he reconstructed the jumble of rough notes left by Cotes at his sudden death.

In the simple situation shown in figure 4.6 it is easy to verify Cotes' theorem by direct calculation of the distances involved, but what if n were very much larger (say, 924)? We don't want to calculate 924 distances! We want a *general* proof, valid for all n. Here's how to do it. First, with no loss of generality we can always place the center of the circle at the origin of the coordinate axes, and imagine that P is on the real axis at $z = a$, as shown in figure 4.6. Then, simply notice that the locations of the vertices of the regular n-gon are the solutions to the cyclotomic equation $z^n - r^n = 0$ discussed in the previous chapter. Thus, with z_K denoting the location of the kth vertex, we can write this equation in factored form as

$$(z - z_1)(z - z_2)(z - z_3) \cdots (z - z_n) = z^n - r^n.$$

If we take the absolute value of both sides of this expression, and use the fact that the absolute value of a product is the product of the absolute values—a fact easy to verify by direct evaluation—then

$$|z - z_1| \, |z - z_2| \, |z - z_3| \cdots |z - z_n| = |z^n - r^n|.$$

From assertion 2, the left-hand side is just the product of the distances from a point at arbitrary location z to all the vertices. Now, in particular, let us take $z = a$, the location of point P. Since P is on the real axis, then z is real and $z^n = a^n$. Also, by its very definition as the value of the radius, we know r is real, and so $z^n - r^n$ is real, too. Thus, since an absolute value is always positive, we have

$$|a - z_1| \, |a - z_2| \, |a - z_3| \, |\cdots| \, |a - z_n| = \begin{cases} a^n - r^n \text{ if } a > r \\ r^n - a^n \text{ if } a < r. \end{cases}$$

Again, we are done, and all the action happens so fast we almost have to look twice to realize it.

4.3 THE GAMOW PROBLEM

For our next example, consider the following problem taken from George Gamow's wonderful, popularized science book *One Two Three . . . Infinity*. Published in 1947, it is still one of the best of its kind, maybe even *the* best. Gamow, a physicist, took a utilitarian approach to mathematics, one much more like an engineer's than a mathematician's, and in one section he discusses complex numbers. In particular, he invented a charming problem to illustrate the rotational property of $\sqrt{-1}$. Gamow's problem is presented as the story of a "young and adventurous man" who discovers an ancient parchment among his late great-grandfather's papers. On it he reads:

> Sail to _____ North latitude and _____ West longitude where thou wilt find a deserted island. There lieth a large meadow, not pent, on the north shore of the island where standeth a lonely oak and a lonely pine. There thou wilt also see an old gallows on which we once were wont to hang traitors. Start thou from the gallows and walk to the oak counting thy steps. At the oak thou must turn *right* by a right angle and take the same number of steps. Put here a spike in the ground. Now must thou return to the gallows and walk to the pine counting thy steps. At the pine thou must turn *left* by a right angle and see that thou takest the same number of steps, and put another spike into the ground. Dig halfway between the spikes; the treasure is there.

All of this is shown in figure 4.7. To this wonderfully imaginative set of instructions Gamow added two funny footnotes: one to tell us that he has of course omitted the numerical values of longitude and latitude to prevent any of us from tossing his book aside and rushing off to start digging up the treasure, and a second to inform us that he, of course, knows oak and pine trees don't grow on deserted islands but he has altered the real types of trees again to help keep the actual island secret. Gamow must have been quite a funny fellow at a party.

The young man follows the instructions, at least to the point of locating the island, where he sees the oak and the pine trees. But, alas, there is no gallows! Unlike the living trees, the gallows has long since disintegrated in the weather,

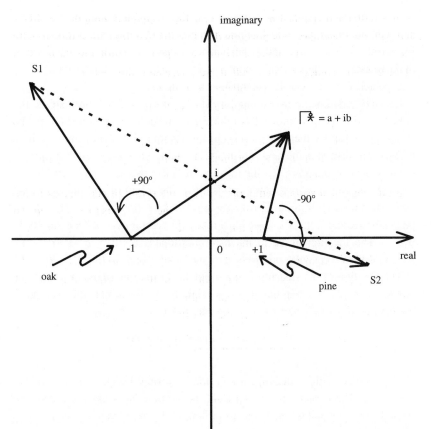

Figure 4.7. Gamow's map.

and not a trace of it or its location remains. Unable to carry out the rest of the instructions (or so he believes), the young man sails away with nary a gold coin or a diamond necklace to show for his troubles. And that's too bad because, as Gamow observes, he could have located the treasure with no difficulty at all, if he had understood complex numbers. Now, while I find Gamow's problem itself charming I am less enthusiastic about his explanation of it. This is one of the very few instances where I would even dare to challenge a thinker of Gamow's intellect, but I think my solution is far clearer. If you want to compare Gamow's with mine, then buy his book—as a true classic of popularization, it is still in print after fifty years. Here's my solution.

Since we don't know where the gallows was, let's just write its location as the general $a + ib$ in the complex plane, as shown in figure 4.7, using a coordinate system with the real axis drawn through the line segment joining the two trees, and with the imaginary axis positioned so that the two trees are symmetrically located at ± 1 (in whatever units of distance we wish). As it turns out, the location of the treasure is independent of both a and b, an astonishing and, I think, utterly unexpected fact that can be established as follows.

Begin by first imagining a temporary shift of the origin of coordinates to the oak. Then the vector pointing from the oak to the gallows is $(a + 1) + ib$. To locate the first spike, then, we must rotate this vector by $+90°$, i.e., multiply it by i. The location of $S1$ in the new coordinates is, then, $-b + i(1 + a)$. Returning to the original coordinates gives the location of $S1$ as $-b - 1 + i(1 + a)$.

Next, imagine a second temporary shift of the origin of coordinates to the pine. Then the vector pointing from the pine to the gallows is $(a - 1) + ib$. To locate the second spike, then, we must rotate this vector by $-90°$, i.e., multiply by $-i$. The location of $S2$ in the new coordinates is, then, $b - i(a - 1)$. Returning to the original coordinates gives the location of $S2$ as $(b + 1) - i(a - 1)$.

The location of the treasure is the midpoint of the line segment joining $S1$ and $S2$ or, as discussed in the previous example, the location is the average of the coordinates of $S1$ and $S2$, i.e., the treasure is at

$$\frac{-(b+1)+i(a+1)+(b+1)-i(a-1)}{2} = i.$$

Our ignorance of the values for a and b is irrelevant, because all the a's and b's cancel. The treasure is located right on the imaginary axis, at a distance from the origin equal to the distances of both of the trees from the origin. Who could have guessed it?

4.4 SOLVING LEONARDO'S RECURRENCE

Recall the generalized Fibonacci recurrence from note 10 of chapter 1, i.e., $u_{n+2} = pu_{n+1} + qu_n$, with u_o and u_1 specified. For Fibonacci's particular problem $p = q = 1$, but we can solve the recurrence (that is, find a formula for u_n as a function of n, alone) for any values of p and q. All it takes is a trick or two, and perhaps just a little bit of complex number arithmetic as well. For example, let us suppose that $p = 4$ and $q = -8$, with $u_o = u_1 = 1$. Then it is simple arithmetic to generate the series:

$$1, 1, -4, -24, -64, -64, \ldots.$$

What I will show you now is how to find a general expression for u_n that will let you calculate any specific term (n given) without having to generate the entire sequence of values up to the term you want.

I will start by guessing $u_n = kz^n$, where both k and z are constants. How do I know that works? Because I've seen it work before, that's how! Now, that really isn't quite the flippant answer you might think it is, as I will justify my "guess" by actually calculating the values of k and z, i.e., I will prove to you that my guess works. There is nothing dishonorable about guessing correct solutions—indeed, great mathematicians and scientists are invariably great guessers—just as long as eventually the guess is verified to work. The next time you encounter a recurrence formula, you can guess the answer too because then *you* will have already seen that it works.

To find the value of z, I'll substitute my guess for u_n into the recurrence. Then,

$$kz^{n+2} = 4kz^{n+1} - 8kz^n,$$

or, upon dividing through by kz^n (notice the k's cancel in every term), we are left with the quadratic $z^2 = 4z - 8$. That is easily solved to give

$$z = \frac{4 \pm \sqrt{16-32}}{2} = 2 \pm i2 = 2^{3/2}e^{\pm i\pi/4}.$$

Note carefully that we have two values for z. That is, both

$$u_{n1} = k_1 2^{3n/2} e^{in\pi/4}$$

and

$$u_{n2} = k_2 2^{3n/2} e^{-in\pi/4}$$

satisfy the recurrence. Note equally carefully that I am not assuming the same value of k for each value of z, i.e., I have put subscripts on the k's to tell them apart. So, which u_n should we use?

The most general solution is to use both, as a sum. Therefore I'll write

$$u_n = u_{n1} + u_{n2} = k_1 2^{3n/2} e^{in\pi/4} + k_2 2^{3n/2} e^{-in\pi/4}.$$

To find the two constants k_1 and k_2, I will next use the so-called *initial conditions*, i.e., the two given values $u_o = u_1 = 1$ that start the recurrence. Thus,

$$n = 0: k_1 + k_2 = 1,$$
$$n = 1: k_1 2^{3/2} e^{i\pi/4} + k_2 2^{3/2} e^{-i\pi/4} = 1.$$

These two simultaneous equations in two unknowns can be solved by routine algebra to give

$$k_1 = \frac{1}{2} + i\frac{1}{4} = \frac{1}{4}\sqrt{5}e^{i\tan^{-1}(1/2)},$$

$$k_2 = \frac{1}{2} - i\frac{1}{4} = \frac{1}{4}\sqrt{5}e^{-i\tan^{-1}(1/2)}.$$

So we have the rather complex-looking result

$$u_n = 2^{3n/2-2}\sqrt{5}\left[e^{i\left\{(n\pi/4)+\tan^{-1}(1/2)\right\}}+e^{-i\left\{(n\pi/4)+\tan^{-1}(1/2)\right\}}\right],$$

but this can be greatly simplified. Using Euler's identity we have

$$u_n = 2^{3n/2-2}\sqrt{5}\;2\cos\left\{\frac{n\pi}{4}+\tan^{-1}\left(\frac{1}{2}\right)\right\}.$$

Now, recalling the formula $\cos(\alpha + \theta) = \cos(\alpha)\cos(\theta) - \sin(\alpha)\sin(\theta)$ derived in section 3.1, and observing that

$$\cos\left\{\tan^{-1}\left(\frac{1}{2}\right)\right\} = \frac{2}{\sqrt{5}},$$

$$\sin\left\{\tan^{-1}\left(\frac{1}{2}\right)\right\} = \frac{1}{\sqrt{5}},$$

it then quickly follows that

$$u_n = 2^{3n/2-1}\left\{2\cos\left(\frac{n\pi}{4}\right)-\sin\left(\frac{n\pi}{4}\right)\right\},\; n = 0, 1, 2, \ldots.$$

It's easy to see that this does reduce to $u_o = u_1 = 1$ for $n = 0$ and $n = 1$, but does it give all the rest of the values of u_n correctly for $n > 1$? Well, it is a simple matter to compare some values of u_n generated directly from the recurrence to the values generated by the above formula. After a while you'll get tired of making such comparisons because they always agree. For example, if you use the recurrence itself you can calculate from it that $u_{11} = -98{,}304$. The formula gives

$$u_{11} = 2^{33/2-1}\left\{2\cos\left(\frac{11\pi}{4}\right)-\sin\left(\frac{11\pi}{4}\right)\right\}$$

$$= 2^{15}\sqrt{2}\left\{2\cos\left(\frac{3\pi}{4}\right)-\sin\left(\frac{3\pi}{4}\right)\right\}$$

$$= 2^{15}\sqrt{2}\left\{-\frac{2}{\sqrt{2}}-\frac{1}{\sqrt{2}}\right\}=-3\cdot 2^{15} = -98{,}304.$$

The formula works.

If you repeat this analysis for the general recurrence, keeping p and q as literals, you'll find that z is complex only if $p^2 + 4q < 0$. So with $p = q =$

1, as in Leonardo's problem, this inequality is *not* satisfied and in his case z will be real. The solution goes through just as easily (perhaps more easily) as it does for complex z, however, and if you try it for yourself you should be able to show that the solution to $u_{n+2} = u_{n+1} + u_n$ with $u_o = u_1 = 1$ is

$$u_n = \frac{1}{\sqrt{5}}\left[\left(\frac{1+\sqrt{5}}{2}\right)^{n+1} - \left(\frac{1-\sqrt{5}}{2}\right)^{n+1}\right], \; n = 0, 1, 2, \ldots.$$

Such recurrences often occur in engineering and science (they are very common in the classical theory of combinatorial probability), but in Leonardo's time they were brand new. Indeed, Leonardo's recurrence was the first time such a thing had been encountered, and many of his contemporaries must have wondered just what he was doing. Perhaps that explains the nickname he went under during his lifetime ("Fibonacci" wasn't used until centuries after his death), that of Bigollo. The word comes from the Italian *bighellone* for loafer or ne'er-do-well. Did this pejorative arise because others thought Leonardo was studying mathematics—such as his recurrence—of no practical value? If so, it didn't bother Leonardo much, as he used it, too, when referring to himself.

4.5 Imaginary Time in Spacetime Physics

In this, the final example of the chapter, I want to show you something in which $\sqrt{-1}$ plays the central role in an application quite a bit different from the purely mathematical. To read it through, however, you will have to accept a couple of equations from physics as being true, but I will quote so much authority you will almost surely be willing to go along.

So here's the authority. According to Einstein (and who is going to dispute *him?*), two people who are in uniform relative motion with respect to each other will measure both time and space differently. This statement, from special relativity, is certainly mysterious as I've stated it, so let me put it in more specific form. Suppose we have a person who we agree to say is "not moving," or at least is not *accelerating*. To say "not moving" is a tricky thing, as we have to say not moving with respect to what. But then, if we point to the what that we say is our standard for not moving we are really right back where we started from, because someone could ask us not moving with respect to what? To say someone is not accelerating, however, is not so problematical. Then there are no forces acting on that person, and forces can be measured with instrumentation located right on that person—a physicist would say the instrumentation is *local*. This follows from Isaac Newton's famous second law of motion, and Newton is tough to argue with, too.

So, given this person who is not accelerating, let us put him at the origin of a standard x, y, z coordinate system in three-dimensional space and say that he measures the location in space of anything observed, called an *event*, with respect to these three coordinate axes. If this person in fact observes two events in space, at (x_1, y_1, z_1) and at (x_2, y_2, z_2), then using the Pythagorean theorem he would calculate the spatial distance squared between the two events as

$$s^2 = (x_1 - x_2)^2 + (y_1 - y_2)^2 + (z_1 - z_2)^2.$$

In fact, if the two events are very close to each other, we could write the differential distance squared as

$$(ds)^2 = (dx)^2 + (dy)^2 + (dy)^2.$$

This last expression is called the *distance metric* of our three-dimensional space.

The Pythagorean or Euclidean function is the usual distance function or metric, sometimes called distance as the crow flies, but it is not the only one possible. Mathematicians have defined the general properties of *all* distance functions as follows: if A and B are any two points, and if $d(A,B)$ denotes the distance between A and B, then (1) $d(A,B) = d(B,A)$; (2) $d(A,B) = 0$ if and only if $A = B$; and (3) if C is any third point then $d(A,B) \leq d(A,C) + d(C,B)$. The Pythagorean distance function possesses these three properties, but so do many other distance functions.[2]

A physical property of $(ds)^2$ that I want to focus on in this case study is its *invariance* with respect to certain classes of changes in the coordinate system. Suppose we draw a line on a flat sheet of paper, and measure the distance between two points (call them A and B) on the line. This distance obviously does not depend on how we happen to draw the coordinate axes, i.e., the distance between A and B will not change if we arbitrarily translate and/or rotate the axes, as shown in figure 4.8. The individual coordinates of A and B *will* obviously change under such coordinate system transformations, but they will change in such a way that the quantity $(ds)^2$ does *not* change. Thus, if we denote the axes of any such shifted (translated) and/or rotated coordinate system by x' and y', then no matter what the coordinates of A and B are in the x, y system, we must have

$$(ds)^2 = (ds')^2 = (dx')^2 + (dy')^2 + (dz')^2.$$

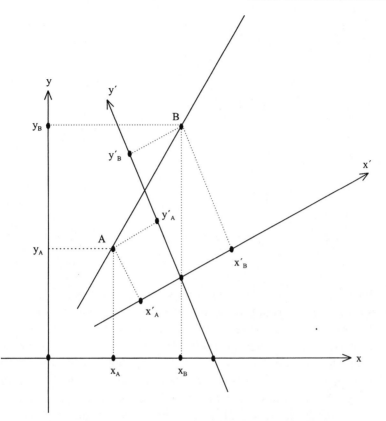

Figure 4.8. The distance between two points does *not* depend on
the coordinate axes.

Now, let me complicate matters just a little bit. In addition to recording the spatial coordinates of an event, i.e., where it is, let's imagine that our nonaccelerating person also records the time, i.e., when the event occurs. He then has *four* numbers, (x, y, z, t), that locate the event in four-dimensional space-time.

That was easy, so let me extend the situation just a little bit more. Suppose we have a second person who moves at constant speed v, measured with respect to the nonaccelerating first person, in the direction along the x-axis of the first person, as shown in figure 4.9. There is no motion of this second person in the other two spatial directions. This is not, of course, the most general case of two people in uniform relative motion, but I am trying to keep things simple here! Now, this second person can reasonably be considered as at the origin of a second coordinate system, which I'll denote as x', y', z'.

99

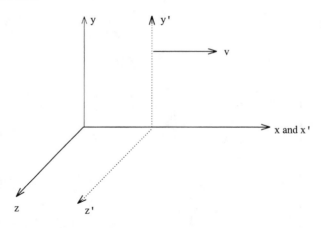

Figure 4.9. Two coordinate systems in uniform, relative motion.

Obviously, $y = y'$, $z = z'$, and, I think, most would agree that $t = t'$, i.e., time "flows" the same for both persons. But what is the connection between x and x'? To answer this question, let us agree to say that we measure time from the instant the two origins in figure 4.9 coincide. That is, at that instant, $t = t' = 0$. Then it is clear, I think, that $x' = x - vt$ because, if both persons see the same event occur at, say, $x = x_o > 0$ at the same time $t_o > 0$, then the moving person will be closer to the event by the distance vt_o.

These elementary considerations, in summary, give us the equations

$$x' = x - vt,$$
$$y' = y,$$
$$z' = z,$$
$$t' = t,$$

equations that are called the *Galilean transformation* from the "stationary" or unprimed coordinate system to the "moving" or primed coordinate system. They are so named to honor the great Italian mathematical physicist Galileo Galilei (1564–1642). They are, however, not correct.

This was the great shock of relativity theory at the start of the twentieth century, in fact, that the "obvious" beliefs that time is the same for everyone, and that space is at most a nearly trivial, uniform shift in coordinates, are simply not true. This is not a book on relativistic physics, however, and so I'll ask you to read up on the physics elsewhere if you are interested,[3] but for our purposes here all you need now is to have the correct transformation equations. They are:

$$x' = \frac{x - vt}{\sqrt{1 - (v/c)^2}},$$

$$y' = y,$$

$$z' = z,$$

$$t' = \frac{t - vx/c^2}{\sqrt{1 - (v/c)^2}},$$

where c is the speed of light, about 186,000 miles per second. Notice that if c were infinity then the correct transformation equations would reduce to the Galilean transformation, i.e., for Galileo, light traveled infinitely fast, which, in most everyday situations, *is* a pretty good approximation.

The correct equations are called the Lorentz (not the Einstein) transformation, after the Dutch physicist Hendrik Antoon Lorentz (1853–1928), who discovered them in 1904 by direct mathematical manipulation of James Clerk Maxwell's equations for the electromagnetic field. It was Einstein, however, who in 1905 showed how to derive the transformation equations from a fundamental reexamination of space and time, without concerning himself about the details of any specific physical laws. *That,* in fact, was Einstein's touchstone, the belief that all physical laws must obey the same coordinate transformation equations—under very general conditions, the Lorentz equations are the only ones that are possible.

With all of this background material behind us, you are ready for the real issue of this example. Rather than talking of the distance between events in three-dimensional space, physicists speak of the *interval* between events in four-dimensional spacetime. How is the interval defined? The obvious answer is simply to extend the three-dimensional definition of distance by adding in the time variable. That is, if we write $c(dt)$ as a sort of natural "time" variable—we have to multiply dt by a speed to get a quantity with space units, to match the units of dx, dy, and dz, and c is the one "natural" speed in physics—then we might reasonably expect the interval ds to be determined by the relation

$$(ds)^2 = (cdt)^2 + (dx)^2 + (dy)^2 + (dz)^2.$$

This so-called *spacetime metric* for the interval may seem natural, but it is *wrong*.

It is wrong because, just as distance is invariant under the coordinate transformations of translation and/or rotation, the interval should be invariant under the Lorentz spacetime transformation. As I'll show you now, however, the above metric is not invariant. To calculate $(ds')^2$, we need to first calculate

$(cdt')^2$, $(dx')^2$, $(dy')^2$, and $(dz')^2$. The last two differentials are particularly easy to calculate because of the geometry of the situation we are assuming in this example: $y' = y$ and $z' = z$. Thus, $dy' = dy$ and $dz' = dz$. For the first two differentials, however, we must do just a little more work. For cdt', it greatly helps to think carefully about just what dt' is. It is the total differential change in time in the primed system, and it depends on two unprimed variables, x and t, because

$$t' = \frac{t - vx/c^2}{\sqrt{1 - (v/c)^2}}.$$

From calculus we have a result due to Euler (1734)

$$dt' = \frac{\partial t'}{\partial x} dx + \frac{\partial t'}{\partial t} dt,$$

which simply says that the total differential change of t' is the sum of the partial differential changes with respect to x and t, with each such change calculated while holding the other variable fixed. Thus,

$$dt' = \frac{-v/c^2}{\sqrt{1 - (v/c)^2}} dx + \frac{1}{\sqrt{1 - (v/c)^2}} dt.$$

Similarly,

$$dx' = \frac{\partial x'}{\partial x} dx + \frac{\partial x'}{\partial t} dt$$

and, since

$$x' = \frac{x - vt}{\sqrt{1 - (v/c)^2}},$$

then we have

$$dx' = \frac{1}{\sqrt{1 - (v/c)^2}} dx - \frac{v}{\sqrt{1 - (v/c)^2}} dt.$$

From these two results it immediately follows that

$$(dt')^2 = \frac{\dfrac{v^2}{c^4}(dx)^2 - 2\dfrac{v}{c^2}(dt)(dx) + (dt)^2}{1 - (v/c)^2},$$

$$(dx')^2 = \frac{(dx)^2 - 2v(dt)(dx) + v^2(dt)^2}{1 - (v/c)^2}.$$

Now, suppose (incorrectly, as you'll soon see) that $(ds)^2 = (cdt)^2 + (dx)^2 + (dy)^2 + (dz)^2$ is the spacetime metric. To check this for invariance, I will next calculate $(ds')^2 = (cdt')^2 + (dx')^2 + (dy')^2 + (dz')^2$. This gives

$$(ds')^2 = \frac{\left[\dfrac{v^2}{c^2}(dx)^2 - 2v(dt)(dx) + c^2(dt)^2\right]}{1-(v/c)^2}$$

$$+\frac{\left[(dx)^2 - 2v(dt)(dx) + v^2(dt)^2\right]}{1-(v/c)^2} + (dy)^2 + (dz)^2.$$

This expression for $(ds')^2$ is clearly *not* equal to $(ds)^2$.

So how *should* we define $(ds)^2$ to have invariance? What Einstein discovered is that rather than using $c(dt)$ we should use $\sqrt{-1}c(dt)$. That is,

$$(ds)^2 = -c^2(dt)^2 + (dx)^2 + (dy)^2 + (dz)^2$$

is invariant under the Lorentz transformation.[4] This means that if we compute $(ds')^2$ we'll get precisely the same result, i.e., that

$$(ds)^2 = (ds')^2 = -c^2(dt')^2 + (dx')^2 + (dy')^2 + (dz')^2.$$

Why, you might wonder, does the expression for $(ds')^2$ have primes on everything except for the c, i.e., why don't we also have to write c' as the speed of light in the primed coordinate system? The answer is that you do, but in fact $c' = c$, i.e., just as is the interval, the speed of light is invariant under the Lorentz transformation. In fact, the invariance of the speed of light is one of the assumptions that Einstein built into his special theory of relativity.[5]

To see that this $(ds)^2$ is invariant write the this last expression out as follows:

$$(ds')^2 = \frac{(dx)^2[1 - v^2/c^2] + (dt)^2[v^2 - c^2]}{1-(v/c)^2} + (dy)^2 + (dz)^2$$

$$= \frac{(dx)^2[1 - (v/c)^2] - c^2(dt)^2[1 - (v/c)^2]}{1-(v/c)^2} + (dy)^2 + (dz)^2$$

$$= (dx)^2 - c^2(dt)^2 + (dy)^2 + (dz)^2$$

which *is* equal to $(ds)^2$. That is, to achieve interval invariance in four-dimensional spacetime we must use $-(cdt)^2 = (c\sqrt{-1}dt)^2$, not $(cdt)^2$. Using $\sqrt{-1}c(dt) = c(\sqrt{-1}dt)$, using if you will "imaginary time" (shades of Buée!),[6] gives us the interval invariance we want, but it also makes it pretty clear that time really is fundamentally different from space, a point that many science writers today have muddied in overly simplistic popularizations on relativity theory.

I ended chapter 3 with the remark that the mathematicians of the nineteenth century often found it hard to accept $\sqrt{-1}$. In some cases, I must tell you, some physicists of the twentieth century had an even harder time. In a revealing article criticizing Einstein's and Minkowski's $c\sqrt{-1}$ a National Bureau of Standards physicist admitted that

> $\sqrt{-1}$ has a legitimate application in pure mathematics, where it forms a part of various ingenious devices for handling otherwise intractable situations. It has also a limited value in mathematical physics, as in the theory of fluid motion, but here also only as an essential cog in a mathematical device. In these legitimate cases, having done its work it retires gracefully from the scene.

But then this physicist demonstrated his real view of $\sqrt{-1}$ as having essentially no physical significance when he concluded his essay with the following sarcastic words:

> The criterion for distinguishing sense from nonsense has been lost; our minds are ready to tolerate anything, if it comes from a man of repute and is accompanied by an array of symbols in Clarendon type.
>
> And yet we must not be too hard on $\sqrt{-1}$; it may stand us in good stead on occasion, as is instanced by a tradition of the National Bureau of Standards.
>
> In the early days of the Bureau, when the staff was smaller, and there were no official guides, the staff-members took turns in conducting parties of visitors through the laboratories. On one such occasion the visitors were shown some liquid air, and they asked, "What is this used for?" In those days liquid air had not yet found any practical application, and was merely a scientific curiosity. The guide . . . was rather non-plussed for the moment, but quickly recovered her presence of mind, and replied, "It is used to lubricate the square root of minus one.[7]

This odd anecdote completely misses the point, as $\sqrt{-1}$ has no less physical significance than do 0.107, 2, $\sqrt{10}$, or any other individual number (about which physicists do not usually write sarcastic essays). Some numbers, of course, *do* have obvious physical ties; for example, π, which is the ratio of the circumference of a circle to its diameter. Perhaps $\sqrt{-1}$ has at least as much physical significance as π, in fact, when one recalls the rotational property of $\sqrt{-1}$ discussed in chapter 3.

More Uses of Complex Numbers

5.1 TAKING A SHORTCUT THROUGH HYPERSPACE WITH COMPLEX FUNCTIONS

I ended the last chapter in spacetime, and this next example of using complex numbers remains at least a little bit connected with that part of mathematical physics. Watchers of science fiction movies are well acquainted with the idea of *hyperspace wormholes* as spacetime shortcuts from one point to another, paths that are traversable in less time than it takes light to make the straight line trip; for example, see the 1994 film *Stargate.*[1] For the first example of this chapter, I want to show you that mathematicians long ago gave a hint at how such a thing might happen, and in fact did so before Einstein published his general theory of relativity, the theory from which the physical prediction of wormholes comes.

As shown in any freshman calculus textbook, the differential arc length ds along a curve $y = y(x)$ is

$$ds = \sqrt{(dx)^2 + (dy)^2} = dx\sqrt{1 + \left(\frac{dy}{dx}\right)^2}$$

The arc length from $x = 0$ to $x = \hat{x}$, then, is simply

$$s = \int_0^{\hat{x}} \sqrt{1 + y'^2}\, dx.$$

In 1914 the American mathematician Edward Kasner (1878–1955) showed how, if $y(x)$ is complex valued and not simply purely real, this formula leads to a totally unexpected, nonintuitive, indeed, a downright bizarre result.

Kasner started his analysis[2] by observing that if we are given two points P and Q on the curve $y = y(x)$, then there are two obvious ways to travel from P to Q. First, simply move along the curve itself, through the arc length distance s where, let us say, P is associated with $x = 0$ and Q with $x = \hat{x}$. Alternatively, you could connect P and Q with a straight line, i.e., the *chord* between P and Q, and travel along that segment through a distance c. Intuitively, it is clear that $c < s$. Somewhat less intuitive is that $\lim_{\hat{x} \to 0} s/c \geq 1$, with most people wondering how this *limit* could be different from unity. As Kasner wrote, however, "It is easy . . . to construct exceptions [to the case of unity ratio] in the domain of real functions: by making the curve sufficiently crinkly the limit may become say two, or any assigned number greater than unity." Then Kasner dropped his bombshell.

If, Kasner wrote, one allows $y(x)$ to be a complex-valued function then the limit of arc length to chord length can be *less* than one! The old adage that "a straight line is the shortest path between two points" is not necessarily true for complex-valued curves. I can't draw a complex-valued curve for you on a piece of paper, of course—how would *you* draw $y(x) = x^2 + ix$, for example?—but we can still do the formal calculations. This is in analogy to theoretical physicists who cannot sketch the way a hyperspace wormhole would penetrate spacetime from Earth to Pluto in a path only one hundred feet long, but who nonetheless take such a thing seriously, at least on paper. Don't confuse the nature of $y(x) = x^2 + ix$ with that of $z = x + iy$. In the second case it is z that is complex and the y that is plotted on the vertical axis is real, i.e., we are using the y-axis to record just the imaginary part of z. For $y = x^2 + ix$, the quantity on the vertical axis is complex and I simply do not know how to draw that! Here now is a simple example of what Kasner did.

Starting with $y(x) = x^2 + ix$, the chord length from $(0,0)$ to $(\hat{x},\hat{y})$ is

$$c = \sqrt{\hat{x}^2 + \hat{y}^2} = \sqrt{\hat{x}^2 + \hat{x}^4 + i2\hat{x}^3 - \hat{x}^2} = \hat{x}\sqrt{\hat{x}^2 + i2\hat{x}}.$$

Thus, as $\hat{x} \to 0$, we can approximate c as

$$c = \hat{x}^{3/2}\sqrt{i2}, \; \hat{x} \approx 0,$$

where I have made use of the fact that $\hat{x}^2$ goes to zero much faster than does $\hat{x}$.

The arc length between $(0,0)$ and $(\hat{x},\hat{y})$ is

$$s = \int_0^{\hat{x}} \sqrt{1+\{y'(u)\}^2} \; du = \int_0^{\hat{x}} \sqrt{1+ (2u+i)^2} \; du$$

$$= \int_0^{\hat{x}} \sqrt{4u^2 + i4u} \; du.$$

Again, as $\hat{x} \to 0$ we can approximate s as

$$s \approx \int_0^{\hat{x}} \sqrt{i4u} \; du = \sqrt{2}\sqrt{2i}\int_0^{\hat{x}} \sqrt{u} \; du$$

$$= \sqrt{2}\frac{2}{3}\sqrt{i2}\hat{x}^{3/2}, \; \hat{x} \approx 0.$$

These two results tell us that

$$\lim_{\hat{x}\to 0} \frac{s}{c} = \left\{ \sqrt{2}\,\frac{2}{3}\,\sqrt{i2}\,\hat{x}^{3/2} \right\}\left\{ \frac{1}{\hat{x}^{3/2}\,\sqrt{i2}} \right\} = \frac{2}{3}\sqrt{2}$$

$$= 0.9428\ldots.$$

That is, the arc length distance is almost 6% *shorter* than the straight line distance.

Kasner actually showed quite a bit more than this, with a much more general result being that if $y(x) = m_k x^k + ix$, where m_k is any value (in the example I just gave, with $k = 2$, I used $m_2 = 1$), then

$$\lim_{\hat{x}\to 0} \frac{s}{c} = \frac{2\sqrt{k}}{k+1}, \; k = 1,\, 2,\, 3,\, \ldots.$$

For $k = 1$ the ratio is unity, but as k increases the ratio can be made as small as we like. This is a good exercise for you to try your hand at—it's not really very difficult. What *is* difficult, I think, is "seeing" this amazing result. Good luck at it, and if you come up with a solution, let me know!

5.2 MAXIMUM WALKS IN THE COMPLEX PLANE

For our next case study I will leave spacetime and mathematical hyperspace shortcuts for the more ordinary two-dimensional complex plane, but even there you'll see that we can find lots of excitement. To motivate the analysis, let me ask you to think back on a standard problem in summing geometric series from high school algebra, and an amusing way to pose it. I recall first hearing it this way as a high school sophomore in 1955.

A boy and a girl stand facing each other, separated by a distance of two feet. While the girl remains stationary, the boy takes a series of steps toward the girl. The first step is one foot. The second is one-half of a foot. The third is one-quarter of a foot. And so on. As the teacher explained with a smile to his class of innocent fifteen-year-olds, the boy never actually reaches the girl but, within a finite number of steps, is close enough "for all practical purposes." At that, naturally, we all—or, at least I—blushed at the thought of the shy, chaste kiss we imagined then occurred.

Suppose now we literally add a twist to this hoary problem. Let's put our young fellow at the origin of the complex plane, and have him walk forward along the positive real axis for unit distance. He then pivots on his heels in a CCW way through angle θ and walks forward for one-half a unit of distance. He then pivots again through a CCW rotation of θ and moves forward for one-quarter a unit of distance. He continues doing this for an infinity of equal

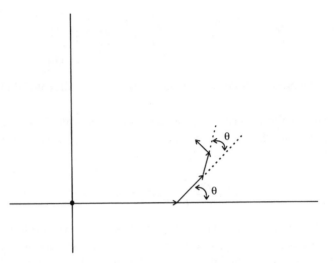

Figure 5.1. A walk in the complex plane.

rotations and ever-decreasing distances, each one-half of the previous distance. Where does he end up in the complex plane, and for what angle θ is he farthest away from the real axis? A sketch of this process is shown in figure 5.1, where I have assumed a value of $\theta < 90°$ simply for the purpose of drawing a clear diagram.

This "walk in the complex plane" is mathematically described by a sum of vectors, i.e., after the $(n + 1)$st step, the vector sum $S(n + 1)$ points to the boy's location in the plane:

$$S(n+1) = 1 + \frac{1}{2}e^{i\theta} + \frac{1}{4}e^{i2\theta} + \cdots + \frac{1}{2^n}e^{in\theta}.$$

The boy's distance from the real axis is the imaginary part of $S(n + 1)$, so what we want to calculate is the value of θ that maximizes the imaginary part of $S(\infty)$. Recognizing the expression for $S(n + 1)$ as a geometric series, with the common term between any two adjacent terms as $\frac{1}{2}e^{i\theta}$, we can multiply through the series by the common factor to get

$$\frac{1}{2}e^{i\theta}S(n+1) = \frac{1}{2}e^{i\theta} + \frac{1}{4}e^{i2\theta} + \cdots + \frac{1}{2^{n+1}}e^{i(n+1)\theta}.$$

Then, subtracting this from the original series, we obtain

$$S(n+1) - \frac{1}{2}e^{i\theta}S(n+1) = 1 - \frac{1}{2^{n+1}}e^{i(n+1)\theta},$$

or

$$S(n+1) = \frac{1 - \frac{1}{2^{n+1}} e^{i(n+1)\theta}}{1 - \frac{1}{2} e^{i\theta}},$$

or, upon letting $n \to \infty$,

$$S(\infty) = \frac{1}{1 - \frac{1}{2} e^{i\theta}} = S_r + iS_i,$$

where S_r and S_i are the real and imaginary parts of $S(\infty)$, respectively. This is not difficult to manipulate to give

$$S(\infty) = S_r + iS_i = \frac{1 - \frac{1}{2}\cos(\theta)}{\frac{5}{4} - \cos(\theta)} + i \frac{\frac{1}{2}\sin(\theta)}{\frac{5}{4} - \cos(\theta)}.$$

Now, to maximize the distance from the real axis, we do the usual thing and set $dS_i/d\theta = 0$. If you do that you will find $\cos(\theta) = 0.8$, or $\theta = 36.87°$, and so the maximum value for $S_i = \frac{2}{3}$, i.e., $S(\infty) = \frac{4}{3} + i\frac{2}{3}$.

Suppose, next, that we don't want to maximize the boy's distance from the real axis but rather his distance from the *origin*. Now we must be somewhat more precise, and specify what is meant by *distance*. For example, do we mean the Pythagorean distance, or the city-block distance I discussed in the last chapter? It makes a difference! (For the problem I just did, however, the answer is the same for either distance function—do you see why?) That is, we can find the θ that maximizes either the Pythagorean distance $\sqrt{S_r^2 + S_i^2}$ or the city-block distance $|S_r| + |S_i|$ and arrive at the results $\theta = 0°$ and $\theta = 15.72°$, respectively. You should go through the calculations and verify these values—without complex exponentials I think these would be quite awkward problems to analyze. In the next example I will show you a problem of ancient origin and of modern significance that is extraordinarily difficult to study without complex exponentials, but with them is extraordinarily pretty.

5.3 KEPLER'S LAWS AND SATELLITE ORBITS

One of the oldest of scientific problems is the so-called N-body problem, inspired by the late-night loneliness of stargazing shepherds and the mysticism of astrologers who earned their livings casting horoscopes. As both of these groups, the earliest astronomers, watched certain specks of light (planets) in the dark sky move across a stationary background of other far more numerous specks of light (stars), the question of how to understand, and even *predict,* such motions was a natural one. In more modern times a justification for an interest in celestial mechanics is the practical problem of predicting the tides and the proper times to launch interplanetary rockets. The positions of the Moon and, to a lesser extent, the Sun, had long been known to be correlated with the variations in the tides, and so there is a natural connection between the two phenomena, one on Earth and the other in the heavens. The military, for one, finds the tides of great interest—when Euler published his 1772 treatise *Theory of the Moon's Motion,* both Russian and British naval navigation experts read it through carefully.

As I will discuss in more detail in this example, Newton's discovery of both the calculus and the inverse-square law of gravitation allowed the motions of two interacting masses to be explained in a scientific, rational way. Newton had begun to think about gravity as early as 1665, when only twenty-two, in connection with the Moon's motion. He used only geometry in his derivation of the inverse-square law of gravity (a law discussed in some detail later in this chapter), not calculus, which he was also in the process of inventing.[3] And later, when he published his gravity research in his 1687 book *Mathematical Principles of Natural Philosophy* (generally called simply the *Principia*), he still restricted his mathematical presentation to geometry. He did this to avoid having the philosophical debates then raging concerning the calculus detract from the physics.

There are, of course, more than just two distinct masses in the universe. The problem of calculating the motion of N gravitationally interacting masses became known as the N-body problem of celestial mechanics, and the myth has spread among most physicists that it remains unsolved for $N \geq 3$. This is true only if one demands closed-form, exact equations. In fact, the Finnish mathematical astronomer Karl F. Sundman (1873–1949) solved the three-body problem during the period 1907–19, and in 1991 a Chinese *student,* Quidong Wang, solved the N-body problem for any N. These solutions are in the form of infinite convergent series, however, which unfortunately converge far too slowly to be of any practical use. Of course, with the development of supercomputers, physicists can now directly calculate the future motions of hundreds, even thousands, of interacting masses, as far into the future as one would like, using Newton's equations of motion and gravity. Solving the N-body problem analytically is no longer of any practical importance.

For $N = 2$ masses, however, exact analytical formulas for the motions have long been known. These derivations are greatly simplified with a couple of approximations. For example, for any particular planet in the solar system it is a good approximation that it exists alone with the Sun. And since the Sun is so much more massive than any of the planets—even Jupiter has less than 0.1% of the Sun's mass—it is a very good approximation that the Sun's motion due to the planet is negligible compared to the planet's motion due to the Sun. To say much more than this requires some interesting mathematics, however, mathematics which is greatly aided by the use of complex numbers. After just a bit of historical discussion, I will show you how Newton literally explained the mystery of the heavens.

The earliest views of the structure of the solar system were geocentric, that the Earth is the fixed center of things, and that everything in the celestial sky that moves revolves in circular paths around our stationary planet. After all, proponents of this view argued, how could the *Earth* move?—everything would blow away in a terrible wind! This objection actually isn't something to sneer at—things do appear that way, at least at first glance. This view was held by the enormously influential Aristotle, for example, who while a great philosopher was a poor scientist. Aristotle can be blamed, for example, for lending support to the erroneous idea that heavy objects fall faster than light objects. He arrived at this conclusion by pure thinking, and apparently never bothered to observe how real objects actually fall. Some disagreed with Aristotle's geocentric view, notably the Greek astronomer Aristarchus, who about 260 B.C. stated that the Sun was the center of all heavenly motions, but their views were dismissed as "obviously" fantastic.

The Earth-as-center view was the one supported in Ptolemy's *Almagest* and it was to hold sway for more than a thousand years. Its hold on the human mind is a bit hard to understand today, because the Earth-centered view has so many observational difficulties. The most obvious problem is that the planets appear, from time to time, to stop temporarily their smooth eastward motions across the night sky and to retrace their paths, i.e., to loop backwards. As early as the fourth century B.C., Greek astronomers were able to "explain" such puzzling retrograde motions, all the while retaining the "perfection" of Aristotle's circular paths, by the mechanism of nested, rotating, invisible crystal spheres centered on the Earth. That is, each of the observed heavenly bodies was imagined to be fixed to one of the inner surfaces of such celestial spheres, which rotated about the Earth at various inclined angles. By adjusting the number of such spheres, and which bodies are stuck on which sphere, the sphere angles, and the sphere rotation rates, the observed motions could then be explained. Of course, as more bodies were more accurately observed more spheres were needed; at one time at least fifty-seven nested spheres were being invoked. How the Greek astronomers explained comets, that swooped

111

in from far out in space and then vanished once again into the night sky, is more of a mystery—they seem not to have been bothered by the question of why such comets did not smash the celestial spheres to pieces.

This situation gradually began to appear ever more ludicrous—would a sane God make such a nightmarish universe?—and resistance to blindly accepting Ptolemy as the last word developed. In 1543, just before he died, the Polish astronomer Nicolaus Copernicus (1473–1543) published his book *On the Revolutions of the Heavenly Bodies,* in which he championed the heliocentric or Sun-centered view. The traditional view is that he held back on publishing his book because he feared religious persecution with such a radical departure from tradition. The Bible, for example, says (in Joshua 10:12–13) that Joshua bade the Sun stand still, not the Earth, an argument used by Martin Luther to refute Copernicus. Modern thought, however, has a more benign view of the Church's position on astronomy at the time of Copernicus' death. During the early stages of the Reformation, the Church had not yet felt sufficiently threatened to have to smash all thought that ran counter to its teachings. Copernicus, in fact, was probably at least as fearful of ridicule from his fellow astronomers (who still took Aristotle as gospel) as he may have been of the Church. Corpernicus was almost certainly correct, however, to be at least cautious of religious fanaticism. That caution is perhaps reflected in the book's presentation of Sun-centered circular orbits as simply a convenient way to calculate the future positions of the planets, and not as physical reality.

That particular bit of philosophical weaseling was not Copernicus' doing, however. The preface to the book was actually written by another, who wrote of Copernicus' work "To be sure, [the author's] hypotheses are not necessarily true; they need not even be probable. It is completely sufficient that they lead to a computation that is in accordance with the astronomical observations." Copernicus' readers thought those were his words, and not until 1854 was it discovered that they were written by another—with whom Copernicus had specifically argued over this very issue—even as the great astronomer lay dying. One subtle bit of weaseling, if that is what it is, *was* due to Copernicus—the very title of his book. *On the Revolutions of the Heavenly Bodies* can be interpreted as implying that only the *other* planets revolve, i.e., move. It could be thought that such a title retains the idea of a stationary Earth.

Copernicus himself certainly held no such view, but he might have been trying to leave himself some "wiggle room" if threatened by the intellectual thugs of the Inquisition. Despite that philosophical squirming, and the dedication of the treatise to Pope Paul III, when the controversy over Galileo's endorsement of the Copernican view erupted the book was placed on the Catholic Church's Index of forbidden books from 1616 until 1757. By 1615 the

Reformation had advanced to the point of a schism in the Church (and the formation of the Protestant movement) and all non-Church teachings were viewed as threats. Copernicus' book had no immediate effect on science, but it did mark the beginning of the end for Ptolemy's geocentric view.

Just three years after Copernicus' death, the Danish astronomer Tycho Brahe (1546–1601) was born, and over the course of his life he made magnificently accurate observations of the orbits of the planets with his unaided eyes—the telescope would not be invented until six years after his death. Tycho, who like Galileo is for some reason always referred to by his first name (pronounced Tee-Ko), made a giant leap backwards, however, by supporting the Earth-centered view. In his last years he had an assistant, the German astronomer Johann Kepler (1571–1630), who inherited the mass of observational data so painstakingly gathered by Tycho. It was Kepler who, after years of studying Tycho's numbers, finally uncovered the rules for how the planets, including Earth, move around the Sun.

In 1609, in his book *New Astronomy,* Kepler announced the first two of his laws of planetary motion: law 1: Each planet moves along an elliptical, not circular orbit with the Sun at one focus, and law 2: The line joining the Sun with any planet sweeps over equal areas in equal intervals of time. The second law tells us, for example, that as a planet approaches the Sun—and so the line joining the two becomes shorter—the planet must travel faster. In 1619, in his book *The Harmony of the Worlds,* Kepler gave his third law: law 3: The ratio of the cube of the semi-major axis of the elliptical orbit to the square of the orbital period is the same for all the planets.

While all the planetary orbits are ellipses, some are more elliptical than others. Recall that if $2a$ and $2b$ are the major and minor axes of an ellipse, respectively, and if $2c = 2\sqrt{a^2 - b^2}$ is the separation of the two foci, then $E = c/a = \sqrt{1 - (b/a)^2}$ is called the *eccentricity* of the ellipse. Obviously, if $c = 0$ then $E = 0$, and the two foci coexist at the center of a circle with diameter $2a = 2b$. The eccentricity of the Earth's orbit is 0.0167, which means $b/a = 0.99986$. Mercury's orbit, on the other hand, has eccentricity 0.2056, which means $b/a = 0.9786$. These orbits are "almost" Copernicus' circles.

Newton deduced both the inverse-square force law for gravity, and the direction of that force, from Kepler's second and third laws. Then, using that deduction, he showed how the first law of elliptical orbits is not an independent statement but rather can be derived as a necessary consequence. Deriving Kepler's elliptical law using only geometry is not easy. In fact, when the physics Nobel prize winner Richard Feynman attempted to prepare a lecture on Kepler's laws for undergraduate students at Caltech, he found he could not follow all of Newton's arguments, which are based on arcane properties of second-degree curves. So Feynman worked out his own geometry-only proof,[4] and it is not easy. As Feynman himself put it, all you need is "infinite intelligence." With these three laws, the *science* of astronomy can be said to have begun. Now, let me show you how all the above can be explained using both Newton's physics and complex exponentials (which is not how Newton, born twelve years after Kepler's death, did it). I'll also show you in the next section, using complex numbers, why the planets can seem to go backward.

Referring to figure 5.2, the complex number *position vector z* from the origin to the moving mass *m* has instantaneous length *r* and makes the instantaneous angle θ with the real axis. "Instantaneous" is an indirect way of saying that *r* and θ are functions of time, i.e., $r = r(t)$ and $\theta = \theta(t)$. That is,

$$|z| = r \text{ and } z = re^{i\theta}.$$

If we write $v = v(t)$ and $a = a(t)$ as the velocity and acceleration, respectively, of the mass *m*, then

$$v = \frac{dz}{dt} = \frac{d}{dt}(re^{i\theta}) = \left[\frac{dr}{dt} + ir\frac{d\theta}{dt}\right]e^{i\theta},$$

$$a = \frac{d^2z}{dt^2} = \frac{dv}{dt} = \left[\frac{d^2r}{dt^2} - r\left(\frac{d\theta}{dt}\right)^2 + i2\frac{d\theta}{dt}\frac{dr}{dt} + ir\frac{d^2\theta}{dt^2}\right]e^{i\theta}.$$

Notice, carefully, that *z*, *v*, and *a* are all vectors, i.e., that all are complex and so have *components*, i.e., projections along both the real and imaginary

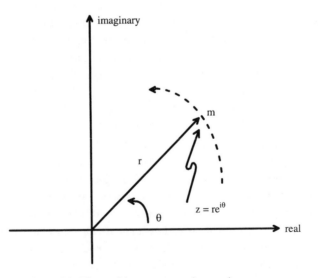

Figure 5.2. The position vector to the moving mass *m*.

axes. Notice, too, that I haven't had to define and introduce the complicating concept of unit vectors that bedevils so many freshman physics students, because the complex exponential automatically takes care of their role for us, of pointing in the right direction.

Now, imagine that there are two masses, M and m, which interact gravitationally according to Newton's inverse-square law. That is, each mass experiences the same force F whose magnitude is given by

$$|F| = G\frac{Mm}{r^2},$$

where G is the so-called universal gravitational constant (first measured in 1798 by the English chemist and physicist Henry Cavendish) and r is the distance between the masses. If the physical extensions of the masses are small compared to their closest approach to each other, then we can generally assume the masses to be *point* masses. If M and m are *spherical* masses with radially symmetric densities, as is approximately true for the Sun and the planets, then we can say r is the distance between the centers of M and m, even if r is small. That this is so was one of the first results Newton obtained through the use of his new calculus. Let us agree to put M at the origin of our coordinate system, and to assume that $M >> m$; for example, the Sun compared to any one of the planets, or the Earth compared to any artificial satellite. This last assumption allows us to ignore the motions of M compared to the motions of m. Finally, let us assume that both M and m are spheres of uniform mass density.

Now, Newton's law of gravitation actually tells us more than just $|F|$. It also says that it is a *radial* or *central* force, i.e., that it acts along the line joining M and m. Further, it is an *attractive* force and so, with M at the origin, the force on m is always directed toward the origin. These facts immediately allow us to write the force vector F on m as

$$F = -G\frac{Mm}{r^2}e^{i\theta}.$$

Inserting the $-e^{i\theta}$ factor, with magnitude unity, gives us the correct value for $|F|$ as well as the correct direction, i.e., it places F along the position vector z but with the opposite direction, which is what we want because gravity is attractive.

Newton's gravitational theory tells us the force exerted by M on m. Newton's laws of motion tell us how m will respond to that force. In particular, $F = ma$ where, as defined earlier, a is the acceleration vector of m. That is,

115

$$a = \frac{d^2 z}{dt^2} = \frac{F}{m} = -G \frac{M}{r^2} e^{i\theta}.$$

$F = ma$ is true only if $m \neq m(t)$. More generally, and as Newton himself stated this law of motion in the *Principia,* force is the time derivative of momentum (or the *motion,* to use his term), which reduces to $F = ma$ for the special case of m constant. Time-varying mass is not an uncommon condition, e.g., a rocket, during launch, is continually losing mass. For the planets orbiting the Sun, of course, it is a pretty good approximation to assume that m is not changing.

We can simplify much of the numerical work in gravitational calculations by referencing everything to the surface of M. There, the gravitational force on a mass m at the surface is precisely what is called the *weight* of the mass m. That force is given by Newton's law of motion as mg, where g is the acceleration of gravity; for example, about 32 feet/second^2 if M is the Earth. But Newton's law of gravity also tells us what this force is, too, and so if R is the distance of the surface of M from the center of M, e.g., about 3,960 miles $= 2.1 \times 10^7$ feet if M is the Earth, then

$$mg = G \frac{mM}{R^2}.$$

Thus

$$G = \frac{rR^2}{M}$$

and so

$$a = -\frac{gR^2}{M} \frac{M}{r^2} e^{i\theta} = -g \frac{R^2}{r^2} e^{i\theta},$$

where it is of course understood that $r > R$, i.e., that the mass m is not inside the mass M. If we simply write $k = gR^2$, then we have

$$a = -\frac{k}{r^2} e^{i\theta}, \quad k = gR^2.$$

But we know a, the acceleration vector of mass m, from our earlier work, and so we have as the differential equation of motion for mass m:

$$\frac{d^2 r}{dt^2} - r \left(\frac{d\theta}{dt} \right)^2 + i2 \frac{d\theta}{dt} \frac{dr}{dt} + ir \frac{d^2\theta}{dt^2} = -\frac{k}{r^2}.$$

This differential equation is perhaps rather intimidating, but actually it is not at all difficult to solve. First, equating the real and imaginary parts of each side, it separates into two simpler differential equations:

$$2\frac{d\theta}{dt}\frac{dr}{dt}+\frac{d^2\theta}{dt^2}=0$$

and

$$\frac{d^2r}{dt^2}-r\left(\frac{d\theta}{dt}\right)^2=-\frac{k}{r^2}.$$

Focusing your attention on the first of these two equations, first multiply through by r to get

$$2r\frac{d\theta}{dt}\frac{dr}{dt}+r^2\frac{d^2\theta}{dt^2}=0,$$

and then notice that this can be written as

$$\frac{d}{dt}\left\{r^2\frac{d\theta}{dt}\right\}=0.$$

You can verify this last statement by simply doing the differentiation. This equation is, of course, integrable by inspection, giving

$$r^2\frac{d\theta}{dt}=c_1,$$

where c_1 is the arbitrary constant of indefinite integration. This result is Kepler's law 2 and it states that the line joining M and m sweeps over equal areas in equal intervals of time. You can see that this is so because if the mass m moves through the differential angle $d\theta$ in differential time dt, then the arc length traveled by m is $rd\theta$. The value of r is essentially constant over the differential time dt, and so the differential area dA swept over by the line of length r from M to m is in the shape of a narrow triangle: $dA = \frac{1}{2}(r)(rd\theta) = \frac{1}{2}r^2d\theta$. Dividing through by dt gives

$$\frac{dA}{dt}=\frac{1}{2}r^2\frac{d\theta}{dt}=\frac{1}{2}c_1 \text{ (= a constant).}$$

Thus, in equal time intervals equal areas will be swept over. If we assume the mass m traverses its orbit in a CCW way, then $d\theta/dt > 0$ and so $c_1 > 0$. I will assume that this is the case from this point on.

Now, combining this last equation with the earlier equation for the acceleration of mass m, we have

117

$$a = \frac{d^2 z}{dt^2} = -\frac{k}{r^2} e^{i\theta} = -k \left(\frac{1}{c_1} \frac{d\theta}{dt} \right) e^{i\theta},$$

which can be integrated once—again, by inspection—to give

$$\frac{dz}{dt} = i \frac{k}{c_1} e^{i\theta} + C,$$

where C is the arbitrary constant of indefinite integration. As you'll soon appreciate, it will be helpful to actually write C as $ic_2 e^{i\theta_0}$, where c_2 and θ_0 are both real, non-negative constants, i.e.,

$$\frac{dz}{dt} = i \frac{k}{c_1} e^{i\theta} + i c_2 e^{i\theta_0}.$$

If you equate this expression for dz/dt to the one I wrote at the very beginning of this mathematical analysis, then we have

$$\left[\frac{dr}{dt} + ir \frac{d\theta}{dt} \right] e^{i\theta} = i \frac{k}{c_1} e^{i\theta} + i c_2 e^{i\theta_0}$$

or, after dividing through by $e^{i\theta}$ which is never zero,

$$\frac{dr}{dt} + ir \frac{d\theta}{dt} = i \frac{k}{c_1} + i c_2 e^{-i(\theta - \theta_0)}.$$

As before, notice that this is really two differential equations, i.e., equating the real and imaginary parts of both sides separately gives

$$\frac{dr}{dt} = c_2 \sin(\theta - \theta_0),$$

and

$$r \frac{d\theta}{dt} = \frac{k}{c_1} + c_2 \cos(\theta - \theta_0).$$

Remembering the mathematical statement of Kepler's area law derived earlier, we have $d\theta/dt = c_1/r^2$ and so the second of the last two equations becomes

$$\frac{c_1}{r} = \frac{k}{c_1} + c_2 \cos(\theta - \theta_0)$$

and thus, at last, we have the orbital equation, in polar form, for the mass m:

$$r = \frac{c_1^2/k}{1 + \dfrac{c_1 c_2}{k}\cos(\theta - \theta_0)}.$$

This equation represents a closed, repeating, i.e., periodic, elliptical orbit with the mass M at the origin, which is also a focus. That is, we have Kepler's law 1 if certain conditions are satisfied. First, as a start to seeing what these conditions are, let me point out the physical significance of θ_0: when $\theta = \theta_0$ the cosine is at its *maximum* value and so r is at its *minimum* value (assuming $c_1 c_2/k \geq 0$, which means $c_2 \geq 0$), i.e., m is at its closest approach to M. This closest-approach distance, called *perigee* if M is the Earth and *perihelion* if M is the Sun, is

$$r_P = \frac{c_1^2/k}{1 + \dfrac{c_1 c_2}{k}}.$$

When $\theta = \theta_0 + \pi$ the cosine is at its *minimum* value and so r is at its *maximum* value, i.e., m is at its *maximum* distance from M, called *apogee* if M is the Earth and *aphelion* if M is the Sun, and

$$r_A = \frac{c_1^2/k}{1 - \dfrac{c_1 c_2}{k}}.$$

We can always orient our coordinate system so as to have the closest approach occur when the orbit crosses the real axis, i.e., we can assume $\theta_o = 0$ with no loss of generality, and I will do that now.

Notice that if we are to have a closed, repeating orbit we must always have r finite and so $c_1 c_2/k < 1$, or else r will, for some values of θ, become arbitrarily large and/or negative. The value of $c_1 c_2/k = E$ is called the eccentricity of the elliptical orbit, and obviously if $E = 0$ the ellipse degenerates into a circle. More precisely, the orbit is a *conic* with eccentricity E: it is an ellipse if $0 < E < 1$, a parabola if $E = 1$, and a branch of an hyperbola if $E > 1$. We know $c_1 > 0$, and so we must have $c_2 = 0$ for zero eccentricity. From the units of c_1 (distance^2/time), c_2 (distance/time), and k (distance^3/time^2), it is clear that E is dimensionless and that c_1^2/k is a distance. To see how I arrived at these various units, simply look back at the equations in which c_1, c_2, and k first appear and make the dimensions balance on each side of the equality. This idea, fundamental to *dimensional analysis,* is a powerful tool for engineers and scientists. We can combine all these observations and summarize what we have by writing

119

$$r = \frac{r_0}{1 + E\cos(\theta)}, \; r_0 = \frac{c_1^2}{k}, \; c_2 \ge 0, \; 0 \le E = \frac{c_1 c_2}{k} < 1.$$

We can now derive Kepler's law 3. To keep things simple, let's suppose $E = 0$ and so we have a circular orbit. Then $r = r_o$ and

$$\frac{d\theta}{dt} = \frac{c_1}{r^2} = \frac{c_1}{r_0^2}.$$

If T is the period of the orbit of m, i.e., the time required for θ to vary from 0 to 2π, then

$$T = \int_0^T dt = \int_0^{2\pi} \frac{r_0^2}{c_1} d\theta = 2\pi \frac{r_0^2}{c_1},$$

or

$$T^2 = 4\pi^2 \frac{r_0^4}{c_1^2} = 4\pi^2 \frac{r_0^4}{r_0 k} = \frac{4\pi^2}{k} r_0^3, \; k = gR^2.$$

That is, for a circular orbit, the square of the orbital period is proportional to the cube of the orbital radius. The proportionality constant k in the equation for T^2 depends only on the mass M (recall the definitions of g and R) and so *any* mass m in a circular orbit with radius r_o will have the same orbital period, e.g., when an American astronaut goes for a space walk while in orbit around the Earth, he moves right along in step with the Shuttle even though his mass is far less than that of the Shuttle. This result is true even if $E \ne 0$, but that more general derivation requires the evaluation of the somewhat more difficult integral

$$\int_0^{2\pi} \frac{d\theta}{\{1 + E\cos(\theta)\}^2}.$$

In chapter 7 I will show you how to evaluate this integral using Cauchy's theory of *contour integration* in the complex plane. If, however, you simply can't wait until then to verify Kepler's third law for $E \ne 0$, then here's the value of the integral for you to play with: $2\pi/(1 - E^2)^{3/2}$.

5.4 WHY AND WHEN THE OTHER PLANETS SOMETIMES APPEAR TO GO BACKWARD

When observed from the Earth, the other planets (ancient Greek astronomers knew of five: Mercury, Venus, Mars, Jupiter, and Saturn) periodically seem, at various times, to reverse direction temporarily in their motions across the night sky. The ancient astronomers could "explain" these mysterious retrograde motions with Ptolemy's crystal spheres, but in fact these motions are simply illusions caused by watching one moving thing (a planet) from another moving thing (the Earth). Kepler knew this, and was the first to explain the illusion using diagrams to illustrate his qualitative reasoning. Complex exponentials, however, make the mathematics of what is happening easy to understand as well. Here's how.[5]

By definition, the Earth has an orbital period of one year. Also by definition, the Earth's average distance from the sun of 93,000,000 miles is one astronomical unit (A.U.). So, with the Sun as the center of our coordinate system, and making the good approximation of a circular orbit, we can write the position vector of the Earth as

$$z_{SE} = e^{i2\pi t}.$$

Notice carefully that this expression also involves the assumption that we have set up our coordinate system such that time $t = 0$ defines an instant when Earth's orbit crosses the positive real axis, as shown in figure 5.3.

Now, let's add a second planet to the problem, one that is at distance a from the Sun and that has orbital period $1/\alpha$ (a and $1/\alpha$ are in the units of A.U.'s and Earth years, respectively). When $\alpha > 1$ this second planet has a shorter

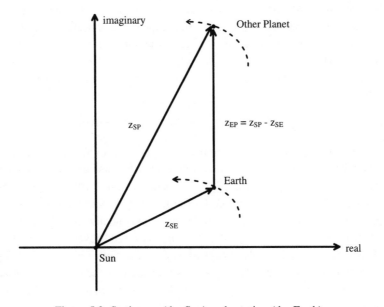

Figure 5.3. Stationary (the Sun) and rotating (the Earth) coordinate system origins.

orbital period than one year, and so must be one of the inner planets (Venus or Mercury). In this case, obviously, $a < 1$. For the outer planets, the situation is the equally obvious $\alpha < 1$ and $a > 1$. These obvious statements all follow mathematically from Kepler's third law, which says

$$\left(\frac{1/\alpha}{1}\right)^2 = \left(\frac{a}{1}\right)^3$$

or $a^3\alpha^2 = 1$. Note carefully that figure 5.3 makes the implicit assumption that the orbits of both the Earth and the other planet are in the same plane. This is not strictly true, but it is almost true. All of the other planets in the solar system do, in fact, have their orbital planes inclined only slightly with respect to the Earth's. With the exception of Pluto—which many astronomers believe is not a true planet but rather is a former satellite of Neptune—the various orbital inclinations do not exceed Mercury's $7°$.

Now, there will in fact be *infinitely* many instances of time when a line joining the Sun to the Earth will also pass through the second planet. We can take one of these instants as time $t = 0$, i.e., we can always pick our coordinate system so that at $t = 0$ the Earth and the second planet are both crossing the positive real axis. Thus, the position vector of the second planet is

$$z_{SP} = ae^{i2\alpha\pi t}.$$

with the Sun as the origin of the coordinate system. By picking α negative we could model a planet that orbits the Sun in the sense opposite that of Earth. In fact, however, all the planets orbit in the same sense, and so it is generally true that $\alpha > 0$.

Referenced to the Earth (the geocentric view) the position vector of the other planet is

$$z_{EP} = z_{SP} - z_{SE} = ae^{i2\alpha\pi t} - e^{i2\pi t} = r(t)e^{i\theta(t)}.$$

The orbital radii of the Earth and the other planet from the Sun are constants, given the approximation of circular orbits, but of course the distance $r(t)$ from the Earth to the other planet will be a function of time. This varying distance manifests itself to observers on the Earth as a time-varying brightness. The planet will also appear to move across the sky in one direction if $\theta(t)$ becomes either ever larger or ever smaller. But if $\theta(t)$ should reverse its direction of change, i.e., if $d\theta/dt$ should change its sign, then the other planet would appear at that instant to reverse direction in its travel across the sky. And that is the retrograde motion we are interested in. So let us see what we can learn about $d\theta/dt$.

We have

$$\ln(z_{EP}) = \ln(r) + i\theta$$

and so

$$\frac{d}{dt}\{\ln(z_{EP})\} = \frac{1}{r}\frac{dr}{dt} + i\frac{d\theta}{dt} = \frac{1}{z_{EP}}\frac{dz_{EP}}{dt}.$$

Thus, $d\theta/dt$ is the imaginary part of $(1/z_{EP})(dz_{EP}/dt)$. But

$$\frac{1}{z_{EP}}\frac{dz_{EP}}{dt} = \frac{ai2\alpha\pi\,e^{i2\alpha\pi t} - i\,2\pi e^{i2\pi t}}{a\,e^{i2\alpha\pi t} - e^{i2\pi t}},$$

which becomes, after a little algebra,

$$\frac{1}{z_{EP}}\frac{dz_{EP}}{dt} = \frac{i2\alpha a^2\pi - i2\alpha a\pi\,e^{-i2\pi t(1-\alpha)} - i\,2\pi ae^{i2\pi t(1-\alpha)} + i2\pi}{1 + a^2 - 2a\cos\{2\pi(1-\alpha)t\}}.$$

If we expand the right-hand side and keep only the imaginary part, then we have

$$\frac{d\theta}{dt} = 2\pi\frac{1 + \alpha a^2 - \alpha a\cos\{2\pi(1-\alpha)t\} - a\cos\{2\pi(1-\alpha)t\}}{1 + a^2 - 2a\cos\{2\pi(1-\alpha)t\}}.$$

We can make this look a little "cleaner" by recalling the earlier statement from Kepler's third law which tells that $a(\alpha a)^2 = 1$, or that

$$\alpha a = \frac{1}{\sqrt{a}}.$$

Substituting this, we arrive at

$$\frac{d\theta}{dt} = 2\pi\frac{1 + \sqrt{a} - (a + 1/\sqrt{a})\cos\}2\pi(1-\alpha)t\}}{1 + a^2 - 2a\cos\{2\pi(1-\alpha)t\}}.$$

Notice that the denominator is always positive for all positive $a \neq 1$ (can you prove this?) and so $d\theta/dt$ changes sign if and only if the numerator changes its sign. That happens at the instants when the numerator equals zero, i.e., $d\theta/dt$ changes sign at the solutions to

$$1 + \sqrt{a} - \left(a + \frac{1}{\sqrt{a}}\right)\cos\{2\pi(1-\alpha)t\} = 0,$$

in other words, at the solutions to

$$\cos\{2\pi(1-\alpha)t\} = \frac{a + \sqrt{a}}{1 + a\sqrt{a}}.$$

123

For any $a > 0$ the right-hand side is less than or equal to one (can you prove this?), and so there are always real solutions to this equation, which means that all of the planets will exhibit retrograde motion when observed from the Earth. This last equation allows us to calculate when this will happen, and for how long as well.

For example, consider the case of Mars. Mars has an orbital period of 687 Earth days and an average distance from the Sun of 141,600,000 miles, and so the values of α and a for that planet are 0.53 and 1.52, respectively. The above equation then becomes

$$\cos\{2\pi(0.47)t\} = 0.958,$$

where $t = 0$ means, as you'll remember, that Mars and the Earth are on a common line drawn from the Sun. That is, at $t = 0$ the Earth is between the Sun and Mars, a situation called *opposition* because the Sun and Mars are on the opposite sides of the Earth. The first solution to the equation is $t = 0.0985$ Earth years $= 36$ days, which means from $t = 0$ days to $t = 36$ days Mars will appear to move in one direction, and then after 36 days it will reverse its motion. Mars will continue to move in the new direction until the next solution, at $t = 2.0295$ Earth years $= 741$ days, at which time Mars will again reverse its direction. It is not difficult to confirm that it is during the shorter time period that $d\theta/dt < 0$, i.e., that that period is the period of retrograde motion. By the symmetry of the cosine function, it is clear that Mars would have been in retrograde motion for 36 days *before* $t = 0$, as well. And so on, i.e., at times separated by somewhat longer than two years, Mars will, according to the previous calculations, periodically display retrograde motion for a duration of about 72 days.

How well does the calculation done above agree with the actual observed motion of Mars? In one college astronomy text[6] that I consulted, after wondering about this question myself, I found an illustration showing the orbit of Mars in the summer and fall of 1988. During that period Mars exhibited retrograde motion between August 26 and October 30, a total of 66 days compared to the 72 calculated here. How might we improve the agreement? The next obvious complication to introduce into this analysis is to use the actual elliptical, not circular, orbits for both Mars and the Earth. Such a more sophisitcated model should result in intervals between the episodes of retrograde motion that are not constant, and the durations of the retrograde motions themselves would not be of equal length. Both of these complicated behaviors are, in fact, observed to occur.

Now, take a look back at the last two sections and I think you'll be astonished, as I am continually, at how much has so easily come from a use of complex exponentials.

5.5 COMPLEX NUMBERS IN ELECTRICAL ENGINEERING

NOTE: In this example, and *only* in this example, I will not use $i = \sqrt{-1}$. Rather, because I will be using the symbol i (with various subscripts) to denote various electrical currents, I will be writing $j = \sqrt{-1}$. This is in accordance with standard electrical engineering practice and so that is what I will do here.

I want to show you now a problem that was a great puzzle until quite late in the nineteenth century. It interested an intellect as elevated as that of Lord Rayleigh, the physics Nobel prize winner in 1904. Rayleigh, whose given name was John William Strutt (1842–1919), was one of the very big men of English science from 1870 until his death half a century later, and his genius cut across practically every aspect of the physics of his day. And yet, in the early days of the study of alternating electrical currents—the well-known "ac"—he ran into a situation that, in his word, was "peculiar." With complex exponentials, however, Rayleigh's puzzle is not too hard to resolve—and that was exactly how Rayleigh himself was able to explain matters.

First, however, a little survey of what you need to know here about electricity. The popular view of electricity two hundred years ago, and a not uncommon one today, was that of a fluid of some mysterious nature, flowing along a "pipe" of wire. Oliver Lodge (1851–1940), one of Rayleigh's English contemporaries, derisively called this the "drainpipe theory" of electricity but, in fact, for constant electrical currents —the well-known "d-c"—flowing in purely resistive circuits (I'll define what they are, soon) the fluid view is enough to get by on. For ac, however, it simply won't do—and that was Lodge's point.

For certain kinds of materials, such as metals and carbon, it is found experimentally that the rate at which electrical charge is transported through the material is directly proportional to the applied voltage. Indeed, the rate *is* the current, as I'll elaborate soon. This statement is called Ohm's law, after the German physicist Georg Ohm (1789–1854) who first published it in 1827. Ohm was a high school teacher who hoped to use his discovery to get a university appointment. This eventually did happen, but not until after a great deal of debate over the merits of his work. Today, of course, every youngster over the age of nine who tinkers with electricity knows Ohm's law and thinks it is obvious. Well, it wasn't always so. It really is not even a *law,* as is the conservation of energy, for example, because it applies only to a limited number of materials.

At this point you might well be thinking to yourself that all this is interesting but still, it might make just a bit more sense to know exactly what *charge* and *voltage* are. Charge is the easier of the two. Just as mass is the property of matter that we say is the basis for gravitational effects (such as falling apples), electrical charge is the property of matter that we say is the basis for electrical effects (such as sparks). Electrical charge is quantized, appearing as integer multiples of the tiny quantum of charge that is the electronic charge: the charge on one electron is 1.6×10^{-19} coulombs, named in honor of the French physicist Charles Coulomb (1736–1806) who discovered Coulomb's law (1785), which describes the nongravitational force between two electrically charged pieces of matter.

Coulomb's force law is analogous to Newton's inverse-square law for gravity. One simply substitutes q and Q for m and M, and uses a different constant in place of the gravitational constant G. As with Newton's law, the force is along the line joining the two charges, but the one great difference is that while masses are always positive (the gravity force is always attractive), electrical charges can be either negative or positive and so the electrical force can be either attractive (the product $qQ < 0$) or repulsive ($qQ > 0$). Another implication of the fact that electrical charges come in both signs is that even matter electrically neutral in the large is still chock-full of electricity—equal amounts of positive and negative electricity that mutually cancel each other's effects.

The movement of charge creates a *current*. If there is charge moving through a conductor—in metals the charge carriers are electrons, present in truly enormous quantity—then a current of one *ampere* is said to be flowing through a cross-section of the conductor if one coulomb of charge passes through the cross-section each second. The unit of electrical current is named after the French mathematical physicist Andre-Marie Ampere (1775–1836), who studied the magnetic fields generated by electrical currents, which play a big role in this example as you will soon see. One ampere is neither a big current nor a small one. The electrical currents in common gadgets like radios and home computers are typically less than one ampere, while the current supplied by an auto battery to the starter motor during the few seconds of cranking can easily exceed several hundred amperes.

Now, a central question. What makes charge move? The answer is an electric field, a concept I'll pursue no further except to say that such fields are produced between the terminals of a battery or an electrical generator. The voltage or *potential difference* between the terminals is what creates an electric field in a conductor, and it is the field that makes the electrons carrying charge in the conductor move (I'll come back to this point in about two paragraphs), and the moving charges *are* the current. If current can be likened to water moving through a drainpipe then voltage is analogous to the pressure. A common flashlight battery produces a potential difference of 1.5 volts, while generators can produce voltages of any value over a wide range, from a few volts to many thousands of volts. When Hoover Dam, on the Colorado River in Nevada near Las Vegas went on line in 1936, for example, it delivered electrical energy to Los Angeles at a potential of 275,000 volts. The unit of voltage is named after the Italian physicist Alessandro Volta (1745–1827), who constructed the first battery in 1800.

The simplest electrical circuits contain only resistors, an electrical component commonly made of carbon in the form of a small cylinder with a wire terminal at each end by which it can be connected to other components. It is mathematically defined to be any device that obeys the relationship $v = iR$, i.e., Ohm's law, where v is the instantaneous voltage difference between the two terminals, i is the instantaneous current in the resistor, and R is the value of the resistor. R is measured in *ohms* if v and i are measured in volts and amperes, respectively.

There are two other two-terminal components that are commonly found in electrical circuits, the capacitor and the inductor. They obey the relationships shown in figure 5.4, where C and L denote the values of those components, in farads and henrys, respectively. The figure also shows the symbols used by electrical engineers when drawing circuit diagrams. The details of how these various components are physically constructed are not important in this example—we'll be interested only in their mathematical definitions. These units are named in honor of the American physicist Joseph Henry (1797–1878) and the English physicist and chemist Michael Faraday (1791–1867). The voltage-current relationships for the various components give us a physical interpretation of what is meant by saying a capacitor has a value of 20 microfarads (20×10^{-6} F), or that an inductor has a value of 500 millihenrys (500×10^{-3} H). The first case means any electrical device that has the property of conducting twenty amperes of current at a particular instant of time if, at that instant, the voltage drop across the device's terminals is changing at the rate of one million volts per second, and such high rates of change are not at all uncommon in certain electronic circuits for very brief intervals of time.

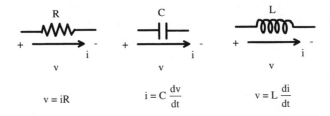

Figure 5.4. The three standard electrical components used in circuits.

The second case means any electrical device that has the property of producing a one volt potential difference between its terminals if the current through it is changing at the rate of two amperes per second.

It is important to notice in figure 5.4 that the current i flows in the direction of the voltage *drop,* i.e., in the direction of going from the + terminal to the − terminal. *The plus/minus symbology can be confusing, as the symbols do not really mean plus and minus.* Rather, the + terminal is simply at a higher voltage than is the − terminal. Both terminals could, in fact, actually be positive when compared to some standard reference, usually called *ground,* which may indeed be the ground of Earth, e.g., +7 volts and +3 volts, and in that case we would mark the +7 volt terminal as + and the +3 volt terminal as −. The voltage drop is +4 volts as we move from + to − (or a voltage rise of +4 volts as we move from − to +).

Now, one final comment about electric fields and the movement of charge. Electric and magnetic fields are vector fields, because at every point in space they have both a magnitude and a direction. Electric fields point in the direction of the voltage drop, i.e., from the + terminal to the − terminal. A mass m carrying a positive electric charge, when placed in an electric field, will experience a force in the direction of the field and so, according to Newton's $F = ma$ formula, will move in that direction. This is consistent with the rule that "like charges repel"; a positive charge at the + terminal will move to the − terminal because it is being repelled. Thus, in electric circuits, the conduction electrons with their negative charge will actually move in the opposite direction, against the field, from the − terminal to the + terminal. That is, the current directions shown in figure 5.4 are opposite to the directions of the actual physical charge carrier motion. No doubt you are thinking this is all the work of madmen—and many beginning electrical engineering students would agree with that assessment —but if you are careful in thinking about what is going on then everything works out.

Now, there are just two things more that I need to tell you about electricity and then you will be ready for Rayleigh's puzzle. In any analysis of an electrical circuit, physicists and electrical engineers use two laws named after the German physicist Gustav Kirchhoff (1824–87). These two so-called Kirchhoff laws are actually the fundamental conservation laws of energy and electric charge—charge and energy can neither be created nor destroyed—when applied to the special case of electric circuits. The laws, illustrated in figure 5.5, are easy to state.

Voltage law: In any circuit, at every instant of time, the sum of all the voltage drops across the components that form any closed loop is zero.

Current law: In any circuit, at every instant of time, the sum of all the currents flowing into any given point is zero. Alternatively, the sum of all the

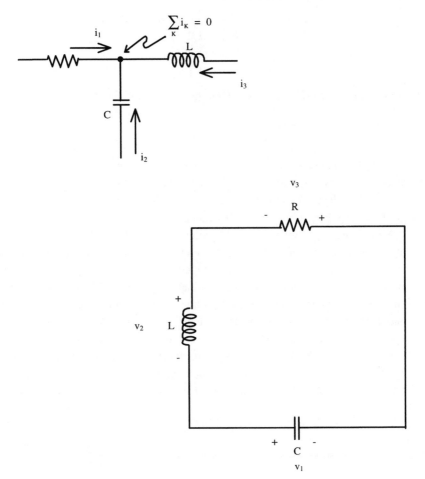

Figure 5.5. Kirchhoff's two circuit laws.

currents flowing into any point equals the sum of all the currents flowing out of that point.

Kirchhoff's laws are very powerful and useful, and will be invaluable in solving Rayleigh's problem. In circuits involving just resistors and batteries, the laws are in agreement with most people's intuition. Intuition and the laws often are *not* in agreement for ac circuits with capacitors and/or inductors present, and in those cases the laws are right and intuition is simply out of luck! For example, consider the circuit shown in figure 5.6. The battery voltage of 1.5 volts appears across both halves of the right-hand side of the circuit. Using Kirchhoff's current law we have $I = I_1 + I_2$. From Ohm's law, for the four resistors, we have

$$V_{ab} = \text{voltage drop from } a \text{ to } b = 0.25I_1,$$
$$V_{bd} = 1.25I_1,$$
$$V_{ac} = 1.25I_2,$$
$$V_{cd} = 1.75I_2.$$

And finally, from Kirchhoff's voltage law, we have

$$1.5 = V_{ab} + V_{bd} = 1.5I_1$$
$$1.5 = V_{ac} + V_{cd} = 3I_2.$$

So I_1 = 1 ampere and I_2 = 0.5 ampere, and thus the battery current is I = 1.5 amperes. It may seem silly to you for me to say this, but notice that I_1 and I_2 are *both* less than I. Well, you might say, of course that must be so because we got I by *adding* I_1 and I_2. Hold that thought.

Now, at last, you are ready for Rayleigh's puzzle. It involves an extension of the simple circuit of figure 5.6 to the one shown in figure 5.7, where I have included all three kinds of electrical components and replaced the battery with an ac voltage generator operating at radian frequency ω. That is, ω has the units of radians per second. This means that if $v(t)$ is a sinusoidal voltage with maximum amplitude of V volts, then we can write $v(t) = V\cos(\omega t)$. The voltage $v(t)$ varies through one complete period, or executes one *cycle* of oscillation, as ωt varies from 0 to 2π. Thus, one cycle of oscillation requires a time interval of $2\pi/\omega$ seconds. Or, alternatively, in one second there will be $1/(2\pi/\omega) = \omega/2\pi$ cycles, a quantity that electrical engineers call the *fre-*

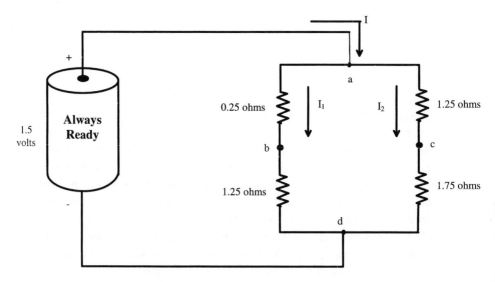

Figure 5.6. A simple resistor circuit.

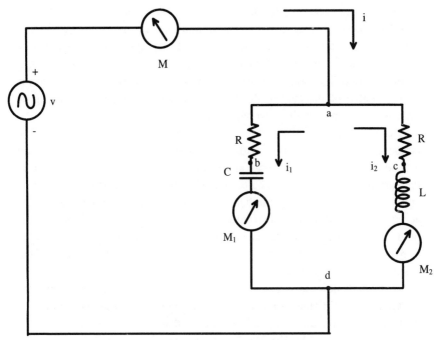

Figure 5.7. Rayleigh's current-splitting puzzle.

qency and denote by f, i.e., $\omega = 2\pi f$. The unit of frequency was for many years the obvious cycle per second. It is now the *hertz,* named for the German physicist Heinrich Hertz (1857–94), who was the experimental discoverer of electromagnetic radiation just a few years before his early death. The frequency of the power lines that distribute electrical energy for ordinary household uses is 60 hertz, while AM broadcast radio uses frequencies between 540,000 hertz and 1,550,000 hertz.

The three circles marked M, M_1, and M_2 in figure 5.7 represent current meters with indicator needles that point to the current value on marked scales beneath the needles. Such meters were invented in 1882 by the French biophysicist Arsène D'Arsonval (1851–1940), who needed a device to measure the tiny electrical currents in animal and human muscle tissue. Called *moving coil D'Arsonval meters,* they utilize the magnetic field generated by a current in a wire. The idea is quite simple in theory, yet delicate to actually implement. A coil of wire is suspended via very-low-friction supports in the magnetic field of a permanent magnet. The current to be measured is passed through the coil, and its magnetic field interacts with the permanent field to produce a force. The whole arrangement is mechanically such that the force

131

produces a torque that rotates the coil, which moves a needle attached to the coil. The force or torque is directly proportional to the instantaneous coil current. If this current is constant, or varies slowly, the coil/needle can vary along with the current, but if the current varies too rapidly then the mechanical inertia of the coil smooths out the variation and so the meter responds to the average value of the current. Such a situation is useless for measuring ac currents, of course, as their average value is zero.

A clever change in the meter design solves that problem, however. The permanent magnet is replaced with an electromagnet, i.e., with a second coil around some iron to concentrate and intensify the magnetic field created by the current to be measured. The current also still passes through the rotating coil, as well. The two magnetic fields now pulsate together in response to the ac current. The instantaneous torque is proportional, just as before, to the product of the fields, i.e., to the *square* of the ac current, and so at "high" frequencies the meter needle deflection is proportional to the average of the *squared* current —which is nonzero even for ac.

What I will now do is calculate, as I did for the simple battery/resistor circuit of figure 5.6, the currents i_1 and i_2 for figure 5.7. What we'll find, and what Rayleigh noticed as far back as 1869, is that while of course $i = i_1 + i_2$ always, as demanded by Kirchhoff's current law, it is nevertheless still possible for the individual currents i_1 and i_2 to both appear *larger* than i! That is, the M_1 and M_2 meter needles can point to larger numbers on their scales than does the M meter needle. Rayleigh himself understood what was going on (indeed, he used complex exponentials almost as I am about to do), but it took a long time before the average electrical engineer on the street felt comfortable with such a nonintuitive phenomenon. The circuit of figure 5.7 is not exactly the one Rayleigh analyzed, nor is my use of complex exponentials exactly like his. But the differences are all minor details.

We can write the following expressions for figure 5.7, using Kirchhoff's laws and the defining relations for resistors, capacitors and inductors:

$$i = i_1 + i_2,$$

$$v = i_2 R + L \frac{di_2}{dt},$$

$$i_1 = C \frac{dv_{bd}}{dt},$$

$$v = i_1 R + v_{bd}.$$

Differentiating the last of these equations, and using the equation for i_1, gives

$$\frac{dv}{dt} = R \frac{di_1}{dt} + \frac{1}{C} i_1.$$

This, together with

$$v = i_2 R + L\frac{di_2}{dt},$$

gives us two differential equations, one each for the currents i_1 and i_2. If we assume v is sinusoidal, then complex exponentials make the solution of these differential equations straightforward.

Suppose that, for each of two different v's (say v_x and v_y), we know the solutions for i_1 (say i_{1x} and i_{1y}). Then, by direct substitution in the differential equation for i_1, it is clear that the solution for the case of $v = v_x + v_y$ is just $i_1 = i_{1x} + i_{1y}$. The same is equally true for i_2. This observation is central to what comes next. In particular, let us write

$$v(t) = 2V\cos(\omega t) = Ve^{j\omega t} + Ve^{-j\omega t},$$

where I have included the factor of 2 to avoid a factor of $\frac{1}{2}$ through all the equations that follow. Since $e^{j\theta}$ represents a vector in the complex plane making angle θ with the real axis, then $e^{j\omega t}$ is a vector making angle ωt which is continually increasing with t, i.e., with time, because $\omega > 0$. That is, $e^{j\omega t}$ is a *rotating vector,* one in fact rotating counterclockwise at positive frequency $\omega/2\pi$ hertz. Similarly, $e^{-j\omega t}$ is a vector rotating clockwise at the same frequency. Electrical engineers often write this as $e^{j(-\omega t)} = e^{j\hat{\omega}t}$, where $\hat{\omega} = -\omega < 0$, i.e., $e^{-j\omega t}$ represents a vector rotating counterclockwise at a negative frequency.

What I will do now is calculate i_1 and i_2 for just the first term of $v(t)$, i.e., for the $Ve^{j\omega t}$ term, and call the results i_1^+ and i_2^+. Then, I will repeat the analysis for the second term, i.e., for the $Ve^{-j\omega t}$ term, and call the results i_1^- and i_2^-. Then, from the observation that opened the previous paragraph, the solutions for $v(t) = 2V\cos(\omega t)$ will be

$$i_1 = i_1^+ + i_1^-,$$
$$i_2 = i_2^+ + i_2^-.$$

It seems reasonable to assume that if $v^+ = Ve^{j\omega t}$ then $i_1^+ = I_1^+ e^{j\omega t}$ and $i_2^+ = I_2^+ e^{j\omega t}$, where I_1^+ and I_2^+ are constants. Substituting v^+, i_1^+, and i_2^+ into the two differential equations gives

$$Vj\omega e^{j\omega t} = RI_1^+ j\omega e^{j\omega t} + \frac{1}{C}I_1^+ e^{j\omega t},$$

$$Ve^{j\omega t} = RI_2^+ e^{j\omega t} + Lj\omega I_2^+ e^{j\omega t},$$

or, upon canceling the common $e^{j\omega t}$ factors that occur in every term,

$$j\omega V = j\omega R\, I_1^+ + \frac{1}{C} I_1^+,$$

$$V = R\, I_2^+ + j\omega L\, I_2^+,$$

Alternatively,

$$I_1^+ = \frac{V}{R - j\dfrac{1}{\omega C}},$$

$$I_2^+ = \frac{V}{R + j\omega L}.$$

Now, if you repeat these steps for $v^- = Ve^{-j\omega t}$, with $i_1^- = I_1^- e^{-j\omega t}$ and $i_2^- = I_2^- e^{-j\omega t}$, you'll quickly find with no difficulty that the results are

$$I_1^- = \frac{V}{R + j\dfrac{1}{\omega C}},$$

$$I_2^- = \frac{V}{R - j\omega L}.$$

Thus, the currents i_1 and i_2 are, for $v(t) = 2V\cos(\omega t)$,

$$i_1 = I_1^+ e^{j\omega t} + I_1^- e^{-j\omega t} = \frac{V}{R - j\dfrac{1}{\omega C}} e^{j\omega t} + \frac{V}{R + j\dfrac{1}{\omega C}} e^{-j\omega t}$$

and

$$i_2 = I_2^+ e^{j\omega t} + I_2^- e^{-j\omega t} = \frac{V}{R + j\omega L} e^{j\omega t} + \frac{V}{R - j\omega L} e^{-j\omega t}.$$

The two expressions for i_1 and i_2 certainly look complex with all the $j = \sqrt{-1}$ factors in them, but in fact they are both purely real. We know this must be so for two reasons, one physical and one mathematical. The physical reason is simply that if we apply the real voltage $v(t) = 2V\cos(\omega t)$ to a circuit made of real (i.e., physically constructable) hardware, then all the resulting voltages and currents must be real, too. What, after all, would a complex current mean? And mathematically, notice that both i_1 and i_2 are the sums of two terms which are complex conjugates. Such a sum is just twice the real part of either term. Since the expressions for i_1 and i_2 are the currents for $v(t) = 2V\cos(\omega t)$, then the currents for $v(t) = V\cos(\omega t)$ are half as much, and so we have

$$i_1 = \text{Re}\left\{ \frac{V}{R - j\dfrac{1}{\omega C}} e^{j\omega t} \right\},$$

$$i_2 = \text{Re}\left\{ \frac{V}{R + j\omega L} e^{j\omega t} \right\}.$$

After a little elementary algebra that I'll leave for you these expressions become

$$i_1 = \frac{V}{R^2 + \left(\dfrac{1}{\omega C}\right)^2} \left[R\cos(\omega t) - \frac{1}{\omega C}\sin(\omega t) \right],$$

$$i_2 = \frac{V}{R^2 + (\omega L)^2} [R\cos(\omega t) + \omega L\sin(\omega t)].$$

For a and b constants, we have the trigonometric identity

$$a\cos(\omega t) + b\sin(\omega t) = \sqrt{a^2 + b^2} \, \cos\left\{ \omega t + \tan^{-1}\left(\frac{b}{a}\right) \right\},$$

and so the currents i_1 and i_2 become [for $v(t) = V\cos(\omega t)$]

$$i_1 = \frac{V}{\sqrt{R^2 + \left(\dfrac{1}{\omega C}\right)^2}} \cos\{\omega t - \tan^{-1}(1/\omega RC)\},$$

$$i_2 = \frac{V}{\sqrt{R^2 + (\omega L)^2}} \cos\{\omega t + \tan^{-1}(\omega L/R)\}.$$

Thus, i_1 and i_2 are sinusoidal currents with squared maximum values I_1^2 and I_2^2, respectively, given by

$$I_1^2 = \frac{V^2}{R^2 + \left(\dfrac{1}{\omega C}\right)^2},$$

$$I_2^2 = \frac{V^2}{R^2 + (\omega L)^2}.$$

The reason for squaring the magnitudes of i_1 and i_2 is that that gives the quantities to which a D'Arsonval current meter responds.

Suppose now that we specify the frequency ω. In particular, suppose $\omega = 1/\sqrt{LC}$. Then it is easy to show that

135

$$i = i + i_2 = \frac{V}{\sqrt{R^2 + \dfrac{L}{C}}} \cos\left\{\omega t - \tan^{-1}\left(\frac{1}{R}\sqrt{\frac{L}{C}}\right)\right\}$$

$$+ \frac{V}{\sqrt{R^2 + \dfrac{L}{C}}} \cos\left\{\omega t + \tan^{-1}\left(\frac{1}{R}\sqrt{\frac{L}{C}}\right)\right\}$$

$$= \frac{2V}{\sqrt{R^2 + \dfrac{L}{C}}} \cos(\omega t)\cos\left\{\tan^{-1}\left(\frac{1}{R}\sqrt{\frac{L}{C}}\right)\right\}$$

$$= \frac{2V}{\sqrt{R^2 + \dfrac{L}{C}}} \cdot \frac{1}{\sqrt{1 + \dfrac{L}{R^2 C}}} \cos(\omega t).$$

So, at the specific frequency of $\omega = 1/\sqrt{LC}$, our earlier results for I_1^2 and I_2^2 become equal, i.e.,

$$I_1^2 = \frac{V^2}{R^2 + \dfrac{L}{C}} = I_2^2,$$

and, from the last calculation, at $\omega = 1/\sqrt{LC}$ the squared maximum value for i, denoted by I^2, is

$$I^2 = \frac{4V^2}{\left(R^2 + \dfrac{L}{C}\right)\left(1 + \dfrac{L}{R^2 C}\right)}$$

From these expressions it is easy to see that $I^2 < I_1^2 = I_2^2$ when $L/(R^2C) > 3$. That is, under that condition (at $\omega = 1/\sqrt{LC}$), meter M in figure 5.7 will point to a *smaller* value than do meters M_1 and M_2, even though it is the splitting of the current in M that forms the currents in M_1 and M_2.

It was a long time before electrical engineers felt comfortable with nonintuitive results like this, which is the reason why ac circuits were thought to be somehow different from battery-powered dc circuitry. *All* circuits obey exactly the same physical and mathematical laws, however, and the breakthrough in understanding this for electrical engineers came in 1893. That year Charles Steinmetz (1865–1923) presented a famous paper at the International Electrical Congress in Chicago titled "Complex Quantities and Their Use in Electrical Engineering." Steinmetz's paper was introduced to the audience with the following words:

We are coming more and more to use these complex quantities instead of using sines and cosines, and we find great advantage in their use for calculating all problems of alternating currents, and throughout the whole range of physics. Anything that is done in this line is of great advantage to science.

Steinmetz, who just five years before had fled Germany, literally in the middle of the night, to avoid arrest for his Socialist beliefs, had almost completed studies for a doctorate in mathematics. His knowledge of advanced mathematics, combined with a genius for practical engineering, was a rare talent in the nineteenth century. He quickly came to the attention of the General Electric Company, where he spent the rest of his life educating electrical engineers to the power of complex numbers in advancing the state of technology. He eventually became known as "the wizard who generated electricity from the square root of minus one." As the final calculation of this chapter let me show you an application of $\sqrt{-1}$ that Steinmetz did not live to see, but which really does generate a special electrical signal from $\sqrt{-1}$.

5.6 A Famous Electronic Circuit That Works Because of $\sqrt{-1}$

The square box in figure 5.8 with the A inside represents an *amplifier*, an electronic gadget that produces an output voltage across its right pair of terminals (v) that is A times larger than the voltage across its left pair of terminals

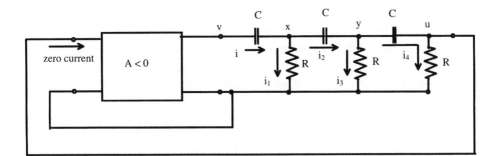

Figure 5.8. An electronic oscillator. The variables v, x, y, and u are voltages, while all the i variables are currents. The bottom terminals at both the input and the output are defined to be the reference with respect to which all other voltages are measured (notice that these two terminals are directly connected together). This is equivalent to saying that the bottom input and output terminals are connected to a *common ground potential* of zero volts.

137

(*u*). That is, *A* is the *voltage gain* of the amplifier. The details of how to construct such a gadget are unimportant to us. All that is important for us to know is that amplifiers do exist and are, in fact, not at all difficult to build. As a basic assumption, I will take it to be true that our amplifier requires so little input current to operate that I can take it to be no current. This ideal assumption is closely approached in actual amplifiers, which can operate with input currents of only billionths of an ampere. To be an amplifier, of course, requires only that $|A| > 1$ but, in its most elementary form, it is easier and cheaper to achieve $A < 0$ rather than $A > 0$. Now, down to business.

The circuit shown in the figure has the rather interesting feature that its input signal comes from the resistor/capacitor arrangement that gets *its* input voltage from the amplifier's *output*. That is, the entire circuit is a closed loop. Can this circuit do anything interesting? This perhaps odd circuit is, in fact, the circuit that started a company that, as I write (1997), has a stock market valuation of over forty billion dollars.

To anticipate physically what I will show you next mathematically, let us suppose a few things. First, imagine that at the amplifier's input there are tiny amounts of voltage present, at any frequency (ω) you want. In fact, imagine that such tiny voltages at all frequencies are all present at once. This will be the case in any circuit because of the reality of electrical *noise,* which is electrical engineering jargon for the random voltages that are always inescapably present. For example, if you tune your television set to a channel where there is no local station broadcasting you will still see rapidly flickering images of black and white dots on the screen, which represent video noise, often called *snow.* Every one of those tiny voltages passes through the amplifier and emerges at the output terminals multiplied by the gain *A;* that is, $|A|$ times larger and inverted because *A* is negative. For example, a tiny input voltage with, let's say, microscopic amplitude μ and a frequency of ω, i.e., *u* $= \mu\cos(\omega t)$, will come out of the amplifier as $v = -|A|\mu\cos(\omega t)$. That voltage then passes through the three resistors and three capacitors and, of course, something will happen to both the amplitude and what electrical engineers call the *phase angle,* i.e., the voltage will emerge as $u = -|A|\mu G(\omega)\cos[\omega t + \phi(\omega)]$, where $G(\omega) > 0$ tells us what the resistors and capacitors do to the amplitude of *v,* and $\phi(\omega)$ is what I called the phase angle. I have written both *G* and ϕ as functions of ω because, most generally, precisely what does happen *is* a function of frequency.

Next, just suppose that there is a particular $\omega = \omega_o$ such that $\phi(\omega_o) = \pi$. Then $u(\omega_o) = -|A|\mu G(\omega_o)\cos(\omega_o t + \pi) = |A|\mu G(\omega_o)\cos(\omega_o t)$. And finally, just suppose we build our amplifier so that $|A|$ is sufficiently large that

$|A|G(\omega_o) = 1$, i.e., that $|A| = 1/G(\omega_o)$. Then $u = \mu\cos(\omega_o t)$. But that is exactly the same voltage we supposed at the input of the amplifier when we started all this "supposing"! That is, the voltage at frequency ω_o is self-sustaining around the loop. All other voltages at all other frequencies will travel around the loop and arrive back out of phase and therefore tend to cancel themselves, and so will not self-sustain. Only at $\omega = \omega_o$ will the self-sustaining process occur. Thus, this circuit will generate a voltage that is a sinusoid at frequency ω_o, i.e., it is a circuit that *oscillates* at a single frequency. It was the invention, at the beginning of the twentieth century, of the electronic vacuum tube that made the construction of such so-called *feedback oscillators* possible, which in turn made modern radio possible.[7]

Of course, the above is really only provocative and not at all analytic. An electrical engineer would demand some mathematics to back it all up—for example, just what *is* ω_o?—and that is what I will do next. The key to understanding what is happening is in the $G(\omega_o)$ of the resistor/capacitor part of the circuit, so let us write down the equations that describe that part. Using the notation of figure 5.8, Kirchhoff's laws, and the voltage-current relations for resistors and capacitors, we have the following set of expressions:

$$i = i_1 + i_2, \quad i = C\frac{d}{dt}(v - x), \quad i_1 = \frac{x}{R}, \quad i_2 = i_3 + i_4,$$

$$i_2 = C\frac{d}{dt}(x - y), \quad i_3 = \frac{y}{R}, \quad i_4 = C\frac{d}{dt}(y - u), \quad i_4 = \frac{u}{R}.$$

With a total of eight equations in nine variables we can solve for the ratio of any two, and the particular ratio of interest for our circuit is u/v. One way to do this is to first manipulate the above equations to eliminate all the variables except for u and v. This is not difficult to do, but it *is* rather detailed, and so I will just give you the answer and encourage you to verify it:

$$\frac{d^3v}{dt^3} = \frac{d^3u}{dt^3} + \frac{6}{RC}\frac{d^2u}{dt^2} + \frac{5}{(RC)^2}\frac{du}{dt} + \frac{1}{(RC)^3}u.$$

To see what this third-order differential equation can tell us, let us next use our old trick of first writing $u^+ = U^+e^{j\omega t}$ and $v^+ = V^+e^{j\omega t}$. If you substitute these expressions into the differential equation you will find that

$$\frac{U^+}{V^+} = \frac{-j\omega^3}{\left(\dfrac{1}{RC}\right)^3 - \dfrac{6\omega^2}{RC} - j\omega\left[\dfrac{5}{(RC)^2} - \omega^2\right]}.$$

Next, write $u^- = U^-e^{-j\omega t}$ and $v^- = V^-e^{-j\omega t}$, substitute, and conclude that

139

$$\frac{U^-}{V^-} = \frac{j\omega^3}{\left(\dfrac{1}{RC}\right)^3 - \dfrac{6\omega^2}{RC} - j\omega\left[\dfrac{5}{(RC)^2} - \omega^2\right]}.$$

Notice that both of these ratios are purely real if the real parts of the two denominators are zero, i.e., if in both cases $\omega = \hat{\omega}$, where

$$\left(\frac{1}{RC}\right)^3 - \frac{6\hat{\omega}^2}{RC} = 0 \text{ or } \hat{\omega} = \frac{1}{RC\sqrt{6}}.$$

For that particular frequency we have

$$\frac{U^+}{V^+} = \frac{-j\hat{\omega}^3}{j\hat{\omega}\left[\dfrac{5}{(RC)^2} - \hat{\omega}^2\right]} = \frac{-\hat{\omega}^2}{\dfrac{5}{(RC)^2} - \hat{\omega}^2}$$

$$= \frac{-\dfrac{1}{6(RC)^2}}{\dfrac{5}{(RC)^2} - \dfrac{1}{6(RC)^2}} = \frac{-1}{30-1} = -\frac{1}{29},$$

and similarly, at $\omega = \hat{\omega}$.

$$\frac{U^-}{V^-} = -\frac{1}{29}.$$

Think carefully about what these results are telling us. At $\omega = \hat{\omega}$ the real and the imaginary parts of u are both precisely equal to $-\frac{1}{29}$th of the real and the imaginary parts of v, respectively. Thus, if the real voltage $2\cos(\hat{\omega}t) = (e^{j\hat{\omega}t} + e^{-j\hat{\omega}t})$ enters the resistor/capacitor circuitry at the left, it will emerge at the right as $-\frac{2}{29}\cos(\hat{\omega}t)$. If the amplifier is designed to have $A = -29$, the voltage output of the amplifier will self-sustain at $\omega = \hat{\omega}$. That is, $\omega_o = \hat{\omega}$. Notice, too, that the value of ω_o is determined by R and/or C, and so by varying either all three identical capacitors or the identical resistors together, via a common mechanical linkage, we can *tune* this circuit to oscillate at any frequency we want over a wide range of frequencies, typically from less than one hertz up to as high as a megahertz or so. For example, the physically reasonable component values of $R = 26,000$ ohms and $C = 0.001$ µF give a frequency of $f_o = \omega_o/2\pi = 2,500$ hertz, a tone right in the middle of the frequency interval that most adults can hear.

This particular feedback oscillator circuit, called a *phase-shift oscillator,* was the basis for the first product developed by two young electrical engineering graduates of Stanford University in the late 1930s, William Hewlett and

David Packard. They used it to make variable-frequency sound generators and sold them to Walt Disney, who used them to create the audio effects in his classic 1940 film *Fantasia*. Today the circuit is so well known that it is often used as the basis for undergraduate laboratory experiments. As I write, in fact, I am teaching two sections of the third-year electrical engineering laboratory at the University of New Hampshire, and one of the experiments asks students to compare theoretical calculations (much those here) with in-lab measurements on a real, operational oscillator. These comparisons routinely agree to within a few percent.

Wizard Mathematics

6.1 LEONHARD EULER

While it is general practice today to date the beginning of the modern theory of complex numbers from the appearance of Wessel's paper, it is a fact that many of the particular properties of $\sqrt{-1}$ were understood long before Wessel. The Swiss genius Leonhard Euler (1707–83), for example, knew of the exponential connection to complex numbers. The son of a rural pastor, he originally trained for the ministry at the University of Basel, receiving, at age seventeen, a graduate degree from the Faculty of Theology. Mathematics, however, soon became his life's passion. He remained a pious man, but there was never any doubt that he was, first, a mathematician.

Nothing could keep him from doing mathematics, not even blindness for the last seventeen years of his life. Euler had a marvelous memory—it was said he knew the *Aeneid* by heart—and so after losing his sight he simply did monstrously difficult calculations in his head. His reputation among his contemporaries was such that he was known as "analysis incarnate." Many years after his death the nineteenth-century French astronomer Dominique Arago said of him "Euler calculated without apparent effort, as men breathe, or as eagles sustain themselves in the wind." When he died he had written more brilliant mathematics than had any other mathematician, and to this day he still holds that record.

While a student at Basel Euler studied with the mathematician John Bernoulli (1667–1748) and, along the way, became friends with two of his sons, Nicolas and Daniel, who were also mathematicians. Several years older than Euler, both soon recognized the younger man's talents, and so when the two Bernoulli boys went off to the Imperial Russian Academy of Sciences in St. Petersburg in 1725 they began to lobby for a spot there for Euler as well. Nicolas died in 1726, but Daniel continued his efforts and in 1727 Euler, too, arrived in Russia. This first of two stays in Russia would see his first great success, and in just a few years (1731) he was named an Academy Professor.

A few days before Euler first set foot in Russia, however, the Tsarina Catherine I (widow of Peter the Great) died and the throne passed to a twelve year old boy. The regency that then ran the country had little sympathy for the intellectual and expensive Academy, which was viewed as a collection of foreign scientists meddling in Russian culture, and Euler no doubt found the place less than totally congenial. When Euler was invited by Frederick the Great of Prussia to leave the Russian Academy and to take up a similar post in the Berlin Academy, he was happy to accept, and there he stayed from 1741 to 1766. He left Berlin because four years earlier Catherine the Great ascended the Russian throne, the intellectual climate there once again became attractive (and Euler was allowed to write his own, generous contract), and his personal relationship with Frederick had deteriorated. And so Euler returned to St. Petersburg. There he remained until his sudden death of a stroke as he sat one evening doing what he loved most—mathematics.

6.2 EULER'S IDENTITY

In a letter dated October 18, 1740 to his one-time teacher John Bernoulli, Euler stated that the solution to the differential equation

$$\frac{d^2 y}{dx^2} + y = 0, \quad y(0) = 2, \text{ and } y'(0) = 0$$

(where the prime notation denotes differentiation) can be written in two ways; namely,

$$y(x) = 2\cos(x),$$
$$y(x) = e^{x\sqrt{-1}} + e^{-x\sqrt{-1}}.$$

The truth of Euler's statement is evident by direct substitution into the differential equation, and the evaluation of each $y(x)$ for the given $x = 0$ conditions. Euler therefore concluded that these two expressions, each apparently so unlike the other, are in fact equal, i.e., that

$$2\cos(x) = e^{ix} + e^{-ix}.$$

It is evident from that same letter that Euler also knew that

$$2i \sin(x) = e^{ix} - e^{-ix}.$$

Just a year after his letter to Bernoulli, Euler wrote another letter (dated December 9, 1741) to the German mathematician Christian Goldbach, in which he observed the near-equality

$$\frac{2^{\sqrt{-1}} + 2^{-\sqrt{-1}}}{2} \approx \frac{10}{13}.$$

Indeed, using the approach sketched in box 3.2 it is easy to show that the left-hand side is cos(ln 2); that and $\frac{10}{13}$ do not begin to differ until the sixth decimal place—only a genius or a quack would notice such a thing, I think, and Euler was no quack!

One mathematician who definitely *did* have something of the quack to him, and who was fascinated by the mystical appearance of the mathematical symbols in Euler's equations, was the Polish-born Józef Maria Hoëné-Wroński (1776–1853), who became a French citizen. He once wrote that the number π is given by the astounding expression

$$\frac{4\infty}{\sqrt{-1}}\left\{(1+\sqrt{-1})^{1/\infty} - (1-\sqrt{-1})^{1/\infty}\right\}.$$

What could he have meant by writing such a thing? Wroński's entry in the *Dictionary of Scientific Biography* uses such words as "psychopathic" and "aberrant," and notes that he had "a troubled and deceived mind," but if one replaces all the infinity symbols with n, writes $(1 \pm i)$ in polar form, i.e., as $\sqrt{2}e^{\pm i\pi/4}$, uses Euler's formulas to expand the complex exponentials, and finally takes the limit as $n \to \infty$, then Wroński's bizarre expression does reduce to 2π. (Not π, as claimed, but perhaps Wroński was thinking of the leftmost infinity as being generated by $\frac{1}{2}n$, not just n—who can say now *what* that odd thinker was thinking?)[1]

Finally, in 1748, Euler published the explicit formula

$$e^{\pm ix} = \cos(x) \pm i \sin(x)$$

in his book *Introductio in Analysis Infinitorum*. To mathematicians, electrical engineers, and physicists this is universally known today as Euler's identity, but as you will soon see he was not the first to either derive it or publish it.

Euler's confidence in this astonishing expression was enhanced by his knowledge of the power series expansion of e^y,

$$e^y = 1 + y + \frac{1}{2!}y^2 + \frac{1}{3!}y^3 + \frac{1}{4!}y^4 + \frac{1}{5!}y^5 + \cdots.$$

If you set $y = ix$, then

$$e^{ix} = 1 + (ix) + \frac{1}{2!}(ix)^2 + \frac{1}{3!}(ix)^3 + \frac{1}{4!}(ix)^4 + \cdots.$$

I am admittedly using this power series expansion for e^y, with y real, in a rather daredevil manner when I substitute an imaginary quantity for y. I am ignoring the question of convergence. I'm doing this because the issue is addressed in great detail in any good book on analysis, where it is shown that the series converges for all z, including complex values, and I simply don't want to turn this book into a textbook. Rest assured, all of the series that I write and treat as convergent in this book *do* converge.

Continuing by collecting real and imaginary parts, we arrive at

$$e^{ix} = \left(1 - \frac{1}{2!}x^2 + \frac{1}{4!}x^4 + \cdots\right) + i\left(x - \frac{1}{3!}x^3 + \frac{1}{5!}x^5 + \cdots\right).$$

But the expressions in the parentheses are the power series expansions for $\cos(x)$ and $\sin(x)$, respectively (known to mathematicians since at least Newton's time), and so Euler's identity is derived in a new way. The sine series, by the way, provides the proof to a statement I asked you just to accept back in section 3.2. Thus we have

$$\frac{\sin(x)}{x} = 1 - \frac{1}{3!}x^2 + \frac{1}{5!}x^4 \cdots$$

and so, with $x = (1/2^n)\,\theta$, we have

$$\lim_{x\to 0} \frac{\sin(x)}{x} = \lim_{n\to\infty} \frac{\sin\left(\dfrac{1}{2^n}\theta\right)}{\left(\dfrac{1}{2^n}\theta\right)} = 1,$$

as claimed.

The power series expansion of e^y was used by both Bernoulli and Euler in some breathtaking calculations. In 1697, for example, John Bernoulli used it to evaluate the mysterious-looking integral $\int_0^1 x^x dx$. Here's how he did it. First, using the trick I used in box 3.2 to calculate $(1 + i)^{1+i}$, he wrote

$$x^x = e^{\ln(x^x)} = e^{x\ln(x)},$$

and then set $y = x \ln(x)$. The power series expansion then gave him

$$\int_0^1 x^x dx = \int_0^1 \left\{\sum_{k=0}^{\infty} \frac{(\ln x)^k}{k!}\right\} dx = \sum_{k=0}^{\infty} \frac{1}{k!}\left\{\int_0^1 (x\ln x)^k dx\right\}.$$

Using integration by parts it is easy to show that

$$\int_0^1 (x\ln x)^k dx = \frac{(-1)^k k!}{(k+1)^{k+1}},$$

a result which is not hard to arrive at if you remember or look up, at the proper time, $\lim_{x\to 0} x \ln x = 0$. From this it immediately follows that

$$\int_0^1 x^x dx = 1 - \frac{1}{2^2} + \frac{1}{3^3} - \frac{1}{4^4} + \frac{1}{5^5} - \cdots = 0.78343 \cdots.$$

6.3 EULER MAKES HIS NAME

Bernoulli's integral was a brilliant calculation, but his former student Euler far surpassed that achievement by using the power series expansion of $\sin(y)$, the imaginary part of e^{iy}, to accomplish what today is still considered a world-class *tour de force*. All he did was solve a problem that had stumped mathematicians for centuries! It also led him to write down a new function, called the *zeta function* today, that is behind the greatest unsolved problem in all of complex number theory; indeed, in all of mathematics. And that was so even before Fermat's last theorem was laid to rest in 1995. Here's what he did.

A mathematical problem of long standing has been the summation of the infinite series of the integer powers of the reciprocals of the positive integers. That is, the evaluation of

$$S_p = \sum_{n=1}^{\infty} \frac{1}{n^p} \text{ for } p = 1, 2, 3, \ldots.$$

The answer for $p = 1$, which results in the so-called *harmonic series,* has been known since about 1350 to *diverge,* a result first shown by the medieval French mathematician and philosopher Nicole Oresme (1320–82).

This conclusion for S_1 surprises most people when they first encounter it, but Oresme's proof of it is beautifully simple. One simply writes S_1 as

$$S_1 = 1 + \frac{1}{2} + \frac{1}{3} + \frac{1}{4} + \frac{1}{5} + \frac{1}{6} + \cdots = 1 + \left(\frac{1}{2}\right) + \left(\frac{1}{3} + \frac{1}{4}\right) + \left(\frac{1}{5} + \frac{1}{6} + \frac{1}{7} + \frac{1}{8}\right) + \cdots$$

and then replaces each term in each grouping on the right with the last (smallest) term in that grouping; notice that this last term will always be of the form $1/2^m$ where m is some integer. This process gives a lower bound on S_1, and so we have

$$S_1 > 1 + \frac{1}{2} + \left(\frac{1}{4} + \frac{1}{4}\right) + \left(\frac{1}{8} + \frac{1}{8} + \frac{1}{8} + \frac{1}{8}\right) + \cdots = 1 + \frac{1}{2} + \frac{1}{2} + \frac{1}{2} + \cdots.$$

That is, we can add $\frac{1}{2}$ to the lower bound on S_1 as many times as we wish, which is just another way of saying that the lower bound itself diverges. But then S_1 must diverge, too.

The divergence is incredibly slow, however. For the partial sum of S_1 to exceed 15, for example, requires well over 1.6 million terms; after 10 *billion* terms the partial sum is only about 23.6, and to reach 100 requires over 1.5×10^{43} terms. Finally, because of its connection to Euler, I should tell you that in 1731 he found that, if $S_1^{(n)}$ is the nth partial sum of S_1, then $\lim_{n \to \infty} \{S_1^{(n)} - \ln(n)\}$ *does* converge, to a number γ now called *Euler's constant*, which is $\gamma = 0.577215664901532 \ldots$. After π and e, γ is perhaps the most important mathematical constant not appearing in elementary arithmetic. In 1735 Euler calculated γ to the fifteen correct decimal places given above, while in modern times it has been calculated to many thousands of places.

There is an elegant way to express γ in terms of S_p, using the power series expansion for $\ln(1 + z)$. This expansion is easily derived for all real z such that $-1 < z < 1$, just as the Danish mathematician Nicolaus Mercator (1619–87) did it in his 1668 book *Logarithmotechnia*. Write $1/(1 + z) = 1 - z + z^2 - z^3 + z^4 - \cdots$, which you can verify by long division. Then integrating both sides gives

$$\ln(1 + z) = z - \frac{1}{2}z^2 + \frac{1}{3}z^3 \cdots + K,$$

where K is the indefinite constant of integration. But since at $z = 0$ we have $\ln(1) = 0$, we must then have $K = 0$, and we are done.

If you now successively substitute the values of $z = 1, \frac{1}{2}, \frac{1}{3}, \frac{1}{4}, \ldots$ into the above expression, you can write the following formulas:

$$1 = \ln(2) + \frac{1}{2} - \frac{1}{3} + \frac{1}{4} - \frac{1}{5} + \cdots,$$

$$\frac{1}{2} = \ln\left(\frac{3}{2}\right) + \frac{1}{2} \cdot \frac{1}{2^2} - \frac{1}{3} \cdot \frac{1}{2^3} + \frac{1}{4} \cdot \frac{1}{2^4} - \frac{1}{5} \cdot \frac{1}{2^5} + \cdots,$$

$$\frac{1}{3} = \ln\left(\frac{4}{3}\right) + \frac{1}{2} \cdot \frac{1}{3^2} - \frac{1}{3} \cdot \frac{1}{3^3} + \frac{1}{4} \cdot \frac{1}{3^4} - \frac{1}{5} \cdot \frac{1}{3^5} + \cdots,$$

$$\frac{1}{n} = \ln\left(\frac{n+1}{n}\right) + \frac{1}{2} \cdot \frac{1}{n^2} - \frac{1}{3} \cdot \frac{1}{n^3} + \frac{1}{4} \cdot \frac{1}{n^4} - \frac{1}{5} \cdot \frac{1}{n^5} + \cdots,$$

If you add these relations together then all the logarithmic terms cancel except one (the sum is said to *telescope*), and you will get

$$\left(1 + \frac{1}{2} + \frac{1}{3} + \cdots + \frac{1}{n}\right) - \ln(n+1) = \frac{1}{2}\left(1 + \frac{1}{2^2} + \frac{1}{3^2} + \cdots + \frac{1}{n^2}\right)$$

$$- \frac{1}{3}\left(1 + \frac{1}{2^3} + \frac{1}{3^3} + \cdots + \frac{1}{n^3}\right)$$

$$+ \frac{1}{4}\left(1 + \frac{1}{2^4} + \frac{1}{3^4} + \cdots + \frac{1}{n^4}\right) - \cdots.$$

So, at last, we have the following exotic double summation (which doesn't converge very rapidly, unfortunately) for γ:

$$\gamma = \lim_{n\to\infty} \left\{ S_1^{(n)} - \ln(n+1) \right\} = \sum_{p=2}^{\infty} \frac{(-1)^p}{p} \left\{ \sum_{n=1}^{\infty} \frac{1}{n^p} \right\}$$

$$= \sum_{p=2}^{\infty} \frac{(-1)^p}{p} S_p.$$

The theoretical importance of γ, and its connection to S_p, explains why the values of S_p are important.

The value of the very next sum after S_1, that is, of S_2, however, baffled all mathematicians who attempted its evaluation, including John Wallis and John Bernoulli. The question of S_2 became widely known in France and England through the 1650 book *Novae Quadraturae Arithmeticae* by the Italian Pietro Mengoli (1625–86), and then from Wallis' *Arithmetica Infinitorum* five years later. Both writers were stumped by the calculation, and they weren't alone. Indeed, the eighteenth-century French historian of mathematics Jean Montucla called the calculation of S_2 "the despair of analysts." One such analyst was Leibniz who, when studying under Christian Huygens as a student in Paris, had declared he could sum any infinite series that converged whose terms followed some rule. But when Leibniz met the English mathematician John Pell (1611–85) in 1673, Pell stopped the exhuberant youth dead in his tracks with S_2. And then, suddenly, Euler solved the problem in 1734 with a derivation using the power series expansion of the sine that is breathtaking in its boldness.

Euler wrote, in essence, the *infinite-degree* polynomial equation

$$f(y) = \frac{\sin(\sqrt{y})}{\sqrt{y}} = 1 - \frac{1}{3!}y + \frac{1}{5!}y^2 - \frac{1}{7!}y^3 + \cdots = 0,$$

and observed that its roots occur when $\sin(\sqrt{y}) = 0$, except that $y = 0$ is not a root because, as shown in the previous section, $\sin(\sqrt{y})/\sqrt{y} = 1$ at $y = 0$, not 0. That is, the roots are at integer multiples of π (excluding 0), i.e., at $\sqrt{y} = n\pi$ where n is any nonzero integer, or at $y = \pi^2, 4\pi^2, 9\pi^2, \ldots$. Now, if we look at just polynomial equations of finite degree, of general form $f(x) = a_n x^n + a_{n-1} x^{n-1} + \cdots + a_1 x + 1 = 0$, with the n roots $r_1, r_2, \ldots, r_n$, then it is plausible (and true) that

$$f(x) = \left(1 - \frac{x}{r_1}\right)\left(1 - \frac{x}{r_2}\right) \cdots \left(1 - \frac{x}{r_n}\right) = 0,$$

since this expression obviously has the same roots and both forms of $f(x)$ are such that $f(0) = 1$.

Using the same reasoning as in Box 1.2, it is clear that the coefficient a_1 can be written as

$$a_1 = -\left(\frac{1}{r_1} + \frac{1}{r_2} + \cdots + \frac{1}{r_n}\right).$$

Euler then reasoned that since this is true for any finite n, no matter how large, it also holds for his infinite-degree polynomial—this is the shaky part of the derivation. Thus, as $a_1 = -1/3! = -1/6$ in the series for $f(y)$, then

$$-\left(\frac{1}{\pi^2} + \frac{1}{4\pi^2} + \frac{1}{9\pi^2} + \cdots\right) = -\frac{1}{6}$$

or

$$\frac{1}{1^2} + \frac{1}{2^2} + \frac{1}{3^2} + \cdots = S_2 = \frac{\pi^2}{6}.$$

That's it! It is no great task to write a simple computer program to calculate the partial sums of S_2 and to watch them approach $1.644934 \ldots$, which is, indeed, $\pi^2/6$. What an incredible result to flow from a knowledge of the power series of the sine function. Euler's method, in fact, allowed him to do much more, to evaluate S_p for all even p. It was this calculation, early in his career at St. Petersburg, that marked Euler as a superstar, and it had much to do with establishing his enormous reputation. For p odd, however, the method fails and those sums remain unknown today. It will take a new Euler, with a new idea, to crack them. Euler tried for many years to calculate S_p for p odd, but in a 1740 paper he could only conjecture that $S_p = N\pi^p$, that N is rational when p is even, and that N will involve $\ln(2)$ when p is odd. In a 1772 paper he did manage to derive[2] a remarkable formula that, while not S_3, is tantalizingly close:

$$\frac{1}{1^3} + \frac{1}{3^3} + \frac{1}{5^3} + \cdots = \frac{\pi^2}{4}\ln(2) + 2\int_0^{\pi/2} x\ln\{\sin(x)\}\ dx.$$

Essentially the only progress made since Euler occurred when the French mathematician Roger Apéry showed (in a paper published in 1979) that S_3, whatever its value may be, is irrational. It is interesting to note that the irrationality of S_2 wasn't known until Legendre proved the irrationality of π^2 in 1794, sixty years after Euler calculated S_2.

6.4 An Unsolved Problem

Once one has the idea of S_p, it is just one more step to considering the sums of the reciprocals of the positive integers raised to a variable power. This Euler did in 1737, writing what is now called the zeta function:

$$\zeta(z) = \sum_{n=1}^{\infty} \frac{1}{n^z} = 1 + \frac{1}{2^z} + \frac{1}{3^z} + \frac{1}{4^z} + \frac{1}{5^z} + \cdots.$$

Euler considered z to be a real integer variable, subject only to the constraint $z > 1$ to insure convergence of the sum. It is clear that the S_p are special cases of the zeta function, e.g., $\zeta(2) = S_2 = \pi^2/6$. More generally, we could think of z as any real number, not necessarily just an integer.

Now, a brief digression. As you almost surely recall from high school mathematics, the primes are all the integers greater than one that have no integer divisors other than one and themselves, i.e., they are not factorable. The sequence of the primes starts with 2, 3, 5, 7, 11, 13, . . . and, as Euclid showed in his *Elements* of about 350 B.C., they are infinite in number although their frequency does decrease as we look at ever larger intervals of numbers. The only fact about the primes that you need to know, to understand what Euler did with $\zeta(z)$, is what is called the *unique factorization theorem*. This theorem states that every positive integer can be written, in exactly one way, as the product of primes. This important result was also known, implicitly, to Euclid. It is easy to prove, a proof can be found in any good book on elementary number theory, and we will take it on faith here.

What Euler did was show that there is an intimate connection between $\zeta(z)$, a continuous function of z, and the primes—which, as integers, are the very signature of discontinuity. It is typical "out of left field, how did he think of that?" Eulerian brilliance.

To begin, multiply through the definition of $\zeta(z)$ by $1/2^z$, to get

$$\frac{1}{2^z}\zeta(z) = \frac{1}{2^z} + \frac{1}{4^z} + \frac{1}{6^z} + \frac{1}{8^z} + \frac{1}{10^z} + \frac{1}{12^z} + \cdots,$$

and so, subtracting this from $\zeta(z)$, we have

$$\left(1 - \frac{1}{2^z}\right)\zeta(z) = 1 + \frac{1}{3^z} + \frac{1}{5^z} + \frac{1}{7^z} + \frac{1}{9^z} + \frac{1}{11^z} \cdots.$$

Next, multiply this last result by $1/3^z$, to get

$$\left(1-\frac{1}{2^z}\right)\frac{1}{3^z}\zeta(z)=1+\frac{1}{3^z}+\frac{1}{9^z}+\frac{1}{15^z}+\frac{1}{21^z}+\frac{1}{27^z}+\frac{1}{33^z}+\cdots,$$

and so, subtracting this from $(1-1/2^z)\zeta(z)$, we have

$$\left(1-\frac{1}{2^z}\right)\left(1-\frac{1}{3^z}\right)\zeta(z)=1+\frac{1}{5^z}+\frac{1}{7^z}+\frac{1}{11^z}+\cdots.$$

Next, multiply this last result by $1/5^z$ and so on . . . well, you see the pattern now, I'm sure.

As we repeat this process over and over, multiplying through our last result by $1/p^z$, where p denotes successive primes, we relentlessly subtract out all the multiples of the primes. You may recognize this as essentially the technique called Eratosthenes' sieve, developed by the third century B.C. Greek mathematician Eratosthenes of Cyrene as an algorithmic procedure for finding all of the primes in the first place. If we imagine doing this multiplication-and-subtraction process for all of the primes, then when we are done—after an infinity of operations, of course—we will have removed every term on the right-hand side of the last expression above exactly once because of the unique factorization theorem. Thus, using Π as the product symbol as in chapter 3,

$$\left\{\prod_{p\text{ prime}}\left(1-\frac{1}{p^z}\right)\right\}\zeta(z)=1$$

or

$$\zeta(z)=\prod_{p\text{ prime}}\left(1-\frac{1}{p^z}\right)^{-1}=\sum_{n=1}^{\infty}\frac{1}{n^z},\quad\text{Re}(z)>1,$$

where $\text{Re}(z)>1$ means the "real part of $z>1$" (to insure convergence of the sum)—this is anticipating the further extension of z from being purely real to being complex. This is the so-called Eulerian product form of the zeta function.

What an incredible result! It is, I think, so unexpected that nobody could possibly have just made it up. It had to be *discovered*. It is typical of so many of Euler's calculations, as it leaves a student almost stunned at the heights some human brains have reached. As an immediate result, it gave Euler an entirely new proof of the infinity of the primes, the first since Euclid's from two thousand years before. It is very short: simply recall that $\zeta(1)=S_1=\infty$. To be consistent with this, however, the above product must have an infinite

151

number of factors, each of which is between 1 and 2, i.e., there must be an infinity of primes.

Euler actually went far beyond this and proved that $\Sigma_{n \text{ prime}} 1/n = \infty$, which is much stronger than merely showing the primes are infinite in number. After all, there are an infinite number of squares, too, and yet, as Euler showed, $S_2 = \Sigma_{n=1}^{\infty} 1/n^2 < \infty$. In some sense, then, there are more primes than there are squares! This result follows from Euler's product formula, again for the case of $z = 1$. Then,

$$\sum_{n=1}^{\infty} \frac{1}{n} = S_1 = \prod_{p \text{ prime}} \left(1 - \frac{1}{p}\right)^{-1}$$

Taking logarithms of both sides, we have (since the log of a product is the sum of the logs)

$$\ln(S_1) = - \sum_{p \text{ prime}} \ln\left(1 - \frac{1}{p}\right).$$

Putting $z = 1/p$ into Mercator's power series expansion of $\ln(1 + z)$, we get

$$\ln\left(1 - \frac{1}{p}\right) = -\frac{1}{p} - \frac{1}{2}\cdot\frac{1}{p^2} - \frac{1}{3}\cdot\frac{1}{p^3} - \cdots,$$

and so

$$\ln(S_1) = \sum_{p \text{ prime}} \left\{\frac{1}{p} + \frac{1}{2}\cdot\frac{1}{p^2} + \frac{1}{3}\cdot\frac{1}{p^3} + \cdots\right\}$$

$$= \sum_{p \text{ prime}} \frac{1}{p} + \text{"finite value."}$$

The finite value follows because the terms beyond $1/p$ can be replaced with larger terms that form a geometric series, which is easily summed using the trick from section 5.2, and from that it is a final, easy step to the finite-value conclusion. Now, since $S_1 = \infty$ then $\ln(S_1) = \infty$ and so $\Sigma_{p \text{ prime}} 1/p = \infty$, too, a result that is, I think, vastly more counterintuitive than is the divergence of the harmonic series, which itself is pretty nonintuitive.

In 1859 the brilliant German mathematician Georg Riemann (1826–66) extended $\zeta(z)$ to complex values of z. He did this in the course of his study of $\pi(x)$, the integer-valued function that is equal to the number of primes not exceeding x, e.g., obviously $\pi(\frac{1}{2}) = 0$ and $\pi(6) = 3$, and not so obviously $\pi(4 \times 10^{16}) = 1,075,292,778,753,150$. Riemann conjectured, but was unable to prove (and nobody else has, either), that all the complex zeros of $\zeta(z)$ are "very likely" of the form $z = \frac{1}{2} + ib$. A *zero* is a value of z such that $\zeta(z) = 0$. That is, what has become known as the *Riemann hypothesis* states that all the complex zeros are on the vertical line in the complex plane, called the *critical line,* with real part equal to $\frac{1}{2}$. (There are an infinity of real zeros at the negative, even integers, i.e., at $z = -2, -4, \ldots$, a result that will be easy to see by the end of this chapter. So easy, in fact, that the real zeros are often called the "trivial zeros.") Because of the Eulerian product form of $\zeta(z)$, which converges for $\mathrm{Re}(z) > 1$, it is clear that $\zeta(z)$ has no complex zeros strictly to the right of the vertical line $\mathrm{Re}(z) = 1$. Otherwise, at least one of the factors in the product would have to be zero, and it is clear that no factor of the form $1 - 1/p^z$ can be zero. Riemann's hypothesis is consistent with this, saying that there are no complex zeros for $\mathrm{Re}(z) > 1$ because all the complex zeros are on the line $\mathrm{Re}(z) = \frac{1}{2}$.

If you have read all this carefully then perhaps you think you've spotted an inconsistency with Riemann's conjecture. How can I talk, you ask, of $z = \frac{1}{2} + ib$ when $\zeta(z)$ is defined only for $\mathrm{Re}(z) > 1$? This valid objection is taken care of by extending the definition of $\zeta(z)$ to all z through a technical process called *analytic continuation.* This is a topic beyond the level of this book, but here is a simple way of getting the spirit of what's involved. Consider the function

$$f(z) = \sum_{n=1}^{\infty} \frac{(-1)^{n+1}}{n^z} = 1 - \frac{1}{2^z} + \frac{1}{3^z} - \frac{1}{4^z} + \frac{1}{5^z} - \cdots,$$

which looks (except for the alternating signs) like $\zeta(z)$. Unlike $\zeta(z)$, however, $f(z)$ *does* converge for $\mathrm{Re}(z) > 0$, precisely because of the alternating signs. Now, for $\mathrm{Re}(z) > 1$, where both $\zeta(z)$ and $f(z)$ are defined, there is a simple connection between $f(z)$ and $\zeta(z)$, namely, that

$$f(z) = \zeta(z) - 2 \sum_{n=1}^{\infty} \frac{1}{(2n)^z},$$

which can be manipulated to the form $f(z) = (1 - 2^{1-z})\,\zeta(z)$. So, when $0 < \mathrm{Re}(z) < 1$, instead of using the divergent series for $\zeta(z)$ we simply calculate $(1 - 2^{1-z})^{-1} f(z)$, which *is* defined for $\mathrm{Re}(z) < 1$, and call the result the value of the *extended Riemann zeta function*—because it was Riemann, not Euler, who first used the zeta symbolism.

In 1914 the English mathematician G. H. Hardy (1877–1947) proved that $\zeta(z)$ has an infinity of complex zeros on the critical line, but that does not prove that *all* the complex zeros are there. More recently, in 1989, it was shown that at least two-fifths of the complex zeros are on the critical line. But that does not say *all* of them are. There is also "very suggestive evidence" from computational studies in 1986, using more than one thousand hours on a supercomputer, that the hypothesis may be true because the first 1.5×10^9 complex zeros above the real axis have actually been calculated and every single one has a real part of $\frac{1}{2}$. But, again, that does not prove *all* the complex zeros are on the critical line.

In fact, most mathematicians attach little significance to such calculations because number theory is full of examples where early numerical work gave results "very suggestive" of the wrong conclusion. A prime example of the danger in reading too much into numerical results can be found in the history of $\pi(x)$. Riemann's interest in the complex zeros of $\zeta(z)$ developed while he was attempting to find a formula for $\pi(x)$. Now, since the time of Gauss it has been known that the logarithmic integral

$$li(x) = \int_2^x \frac{du}{\ln(u)}$$

is an excellent *approximation* to $\pi(x)$, e.g.,

$$\frac{\pi(1,000)}{li(1,000)} = \frac{168}{178} = 0.94382,$$

$$\frac{\pi(100,000)}{li(100,000)} = \frac{9,592}{9,630} = 0.99605,$$

$$\frac{\pi(100,000,000)}{li(100,000,000)} = \frac{5,761,455}{5,762,209} = 0.99987.$$

In an 1849 letter Gauss claimed to have known of this property of $li(x)$ since 1792 or 1793, when he was fifteen. These numerical results are very suggestive of the conjecture $\lim_{x \to \infty} \pi(x)/li(x) = 1$, which is a more precise version of the original statement of the so-called *prime number theorem,* first proven to be true in 1896. Indeed, the theorem was proven that year simultaneously and independently by Hadamard in France and Charles-Joseph de la Vallée-Poussin (1866–1962) in Belgium, each using very advanced arguments from complex function theory applied to the zeta function.

Another famous conjecture based on the same numbers, however, is false. This was the conjecture that $\pi(x)$ is always less than $li(x)$, and that the difference between the two increases as x increases. Gauss, and Riemann too, believed this to be so. In fact, for all values of x for which both $\pi(x)$ and $li(x)$ are known i.e., for x up to 10^{18}, both of these assertions are true. And yet the conjecture is false in general, as was proven in 1914 by Hardy's friend, the English mathematician J. E. Littlewood (1885–1977). Indeed, Littlewood showed that the sign of $\pi(x) - li(x)$, which at first and for so long is negative, actually reverses infinitely often as x increases without bound. So much for "suggestive" numerical results then, and near the end of his life Littlewood stated his personal belief that the Riemann hypothesis itself is also false. To disprove the Riemann hypothesis, of course, one needs to show only that there is at least one complex zero not on the critical line. Nobody has been able to do that, either. As mentioned before, there are no complex zeros of $\zeta(z)$ for $\text{Re}(z) > 1$, and both Hadamard and Vallée-Poussin showed there are no complex zeros on the vertical line $\text{Re}(z) = 1$. To go from no complex zeros on $\text{Re}(z) = 1$ to all complex zeros on $\text{Re}(z) = \frac{1}{2}$, however, has proven, so far, to be too big a gap to jump.

In 1900, at the Second International Congress of mathematicians held in Paris, the great German mathematician David Hilbert (1862–1943) presented a talk titled "Mathematical Problems." During this talk he discussed a number of unsolved problems that he felt represented fruitful directions for future research. These problems become famous challenges, and a mathematician who could solve one could make a career on the resulting acclaim. All were attacked by some of the world's finest mathematical minds. Number eight on the list of twenty-three was the Riemann hypothesis. Hilbert felt that any method that cracked the Riemann hypothesis might, for example, also give insight into the infinity (or not) of the twin primes, i.e., primes that are adjacent odd integers, such as 11 and 13, which is still an open question. To this day, however, the hypothesis remains unbreached, and it is said that Hilbert, when asked what would be the first thing he would say upon awaking after sleeping for five hundred years, replied "I would ask, 'Has somebody proved the Riemann hypothesis?'" It is, in fact, the greatest unsolved problem in mathematics today.

One last story about this problem is particularly amusing. Once, after visiting a mathematician friend in Denmark, Hardy found he would have to take his usual trip home by boat over a particularly rough North Sea. Hoping to divert catastrophe, Hardy quickly, before boarding, wrote and mailed a postcard to his friend declaring, "I have a proof for the Riemann hypothesis!" He was confident, Hardy said later, after reaching England safely, that God would not let him die with the false glory of being remembered for doing what he really had not achieved. Hardy, it should be noted, was a devout atheist in all other aspects of his life.

6.5 EULER'S INFINITE PRODUCT FOR THE SINE

There is one more thing we can do here with the power series expansion of the sine. Instead of expanding $\sin(\sqrt{y})/\sqrt{y}$ into a product, let us do $\sin(y)/y$. Then,

155

$$\frac{\sin(y)}{y} = 1 - \frac{1}{3!}y^2 + \frac{1}{5!}y^4 - \frac{1}{7!}y^6 + \cdots$$

$$= \left(1 - \frac{y^2}{r_1}\right)\left(1 - \frac{y^2}{r_2}\right)\left(1 - \frac{y^2}{r_3}\right)\cdots$$

$$= \left(1 - \frac{y^2}{\pi^2}\right)\left(1 - \frac{y^2}{4\pi^2}\right)\left(1 - \frac{y^2}{9\pi^2}\right)\cdots$$

$$= \prod_{n=1}^{\infty}\left(1 - \frac{y^2}{n^2\pi^2}\right).$$

That is,

$$\sin(y) = y\prod_{n=1}^{\infty}\left(1 - \frac{y^2}{n^2\pi^2}\right),$$

which is Euler's famous product formula for sin(y) that appeared in his 1748 *Introductio in Analysis Infinitorum*. Now, let $y = \pi/2$. Then,

$$1 = \frac{\pi}{2}\prod_{n=1}^{\infty}\left(1 - \frac{\pi^2/4}{n^2\pi^2}\right) = \frac{\pi}{2}\prod_{n=1}^{\infty}\left(1 - \frac{1}{4n^2}\right) = \frac{\pi}{2}\prod_{n=1}^{\infty}\frac{4n^2 - 1}{4n^2}.$$

Or

$$\frac{\pi}{2} = \frac{1}{\displaystyle\prod_{n=1}^{\infty}\frac{(2n-1)(2n+1)}{(2n)(2n)}} = \prod_{n=1}^{\infty}\left(\frac{2n}{2n-1}\right)\left(\frac{2n}{2n+1}\right),$$

or

$$\frac{\pi}{2} = \frac{2}{1}\cdot\frac{2}{3}\cdot\frac{4}{3}\cdot\frac{4}{5}\cdot\frac{6}{5}\cdot\frac{6}{7}\cdots,$$

which is Wallis' product formula for π, mentioned in chapters 2 and 3.

If you take the logarithm of Euler's product formula, which of course transforms the product into a sum, and then differentiate, it is only a line or two of algebra to derive the famous series

$$\frac{1}{\tan(y)} = \cot(y) = \frac{1}{y} + \sum_{n=1}^{\infty}\frac{2y}{y^2 - n^2\pi^2}$$

$$= \frac{1}{y} + \sum_{n=1}^{\infty}\left(\frac{1}{y+n\pi} + \frac{1}{y-n\pi}\right) = \sum_{n=-\infty}^{\infty}\frac{1}{y+n\pi}.$$

Differentiating again, another famous series immediately follows:

$$\frac{1}{\sin^2(y)} = \sum_{n=-\infty}^{\infty} \frac{1}{(y+n\pi)^2}.$$

These and similar series, all of which come from the power series expansion of the sine, were known to Euler as early as 1740. Euler's nifty little trick for expanding $\sin(y)$ as an infinite product can be repeated to show that

$$\cos(y) = \prod_{n=1}^{\infty} \left\{ 1 - \frac{4y^2}{(2n-1)^2 \pi^2} \right\}.$$

As a hint for how to do this, notice that

$$f(y) = \cos(y) = 1 - \frac{1}{2!} y^2 + \frac{1}{4!} y^4 - \cdots$$

has its roots at odd integer multiples of $\frac{1}{2}\pi$.

6.6 BERNOULLI'S CIRCLE

Long before 1748, long before Euler wrote $e^{\pm ix} = \cos(x) \pm i\sin(x)$, the specific calculation of i^i or its equivalent had been accomplished with the use of other, specialized, astonishingly clever calculations.[3] One such calculation was alluded to by Euler himself, in a letter (dated December 10, 1728) that he wrote to John Bernoulli. In that letter Euler cited a result due to Bernoulli himself. In modern terms what Bernoulli had done was to consider the circle of unit radius centered on the origin, i.e., the equation $x^2 + y^2 = 1$, and then he wrote the integral for the area of the first quadrant of the circle. This is, today, a standard problem in freshman calculus, and if we denote the area by A (which is, of course, $\pi/4$) we have

$$A = \int_0^1 y \, dx = \int_0^1 \sqrt{1-x^2} \, dx.$$

Now, change variables to $u = ix$. Then, $x = -iu$ and so $dx = -i \, du$, which leads to

$$A = \int_0^i \sqrt{1-(-iu)^2} \, (-idu) = -i \int_0^i \sqrt{1+u^2} \, du.$$

From integral tables we can write

$$A = \frac{\pi}{4} = -i\left\{\frac{1}{2}u\sqrt{u^2+1} + \frac{1}{2}\ln(u+\sqrt{u^2+1})\right\}\Big|_0^i = -\frac{1}{2}i\ln(i).$$

Thus, $i\ln(i) = -\frac{1}{2}\pi$, from which we immediately have, as before, $i^i = e^{-\pi/2}$, although this last step was not taken by either Bernoulli or Euler at that time.

6.7 THE COUNT COMPUTES i^i

But even this near calculation of i^i was not the first. In 1719 the Italian Giulio Carlo dei Toschi Fagnano (1682–1766) carried out a circle calculation superficially resembling Bernoulli's, only Fagnano worked with the circumference rather than with the area. Born into a noble family that included a pope, Fagnano was made a count in 1721 by Louis XV, and Pope Benedict XIV elevated him to marquis in 1745. A member of both the Royal Society of London and the Berlin Academy of Sciences, he enjoyed an excellent reputation throughout Europe as a creative mathematician. The following calculation helped make that reputation.

What Fagnano did was to begin with the unit circle and to observe that the arc length L, subtended by a central angle of θ, is simply θ. That is, with t as a dummy variable of integration, he wrote

$$L = \theta = \int_0^\theta dt.$$

Then, manipulating the integrand (which could hardly be more elementary as it stands) with simple algebra, we can make it look far more complicated. This certainly seems a silly thing to do, yes, but be patient—there *is* method to the apparent madness. Thus,

$$L = \int_0^\theta \frac{dt/\cos^2(t)}{1/\cos^2(t)} = \int_0^\theta \frac{dt/\cos^2(t)}{\{\sin^2(t)+\cos^2(t)\}/\cos^2(t)}$$

$$= \int_0^\theta \frac{dt/\cos^2(t)}{1+\tan^2(t)}.$$

Next, change variables to $x = \tan(t)$, which says $dx = dt/\cos^2(t)$, and so

$$L = \int_0^{\tan(\theta)} \frac{dx}{1+x^2}.$$

Since L equals one-fourth of the circle's circumference when $\theta = \pi/2$, then

$$\frac{\pi}{2} = \int_0^\infty \frac{dx}{1+x^2}.$$

This result is considered elementary today as the indefinite integral is commonly shown in freshman calculus to be $\tan^{-1}(x)$, and of course $\tan^{-1}(x)\big|_0^\infty = \pi/2$. The weird manipulations I just went through are a way to arrive at the definite integral without having to know the indefinite answer. But that was not the motivation for Fagnano.

The reason Fagnano got the integral for L into this form is that he could now use a clever trick due to John Bernoulli. In 1702, Bernoulli had shown that the integrand in the integral for L can be factored into imaginary components and then split into what is today called a partial fraction expansion. That is,

$$\frac{\pi}{2} = \int_0^\infty \frac{dx}{(x+i)(x-i)} = \int_0^\infty \frac{1}{2i}\left(\frac{1}{x-i} - \frac{1}{x+i}\right)dx.$$

Now, the two resulting integrals on the right are easy to do, as they are the well-known logarithmic integrals. That is,

$$\frac{\pi}{2} = \frac{1}{2i}\Big[\ln(x-i) - \ln(x+i)\Big]\bigg|_0^\infty = \frac{1}{2i}\ln\left(\frac{x-i}{x+i}\right)\bigg|_0^\infty.$$

Bernoulli was utterly fascinated by this approach, as it showed the intimate connection between the arctangent and the logarithmic functions. He gave it the special name of *logarithme imaginaire*. It was in that same year of 1702 that Leibniz said of such factoring of polynomials into imaginary components that they are "an elegant and wonderful resource of divine intellect, an unnatural birth in the realm of thought, almost an amphibium between being and non-being." Bernoulli and Count Fagnano would certainly have agreed with that.

Now, continuing on with the Count's calculation, if we assume we can ignore the i in $x \pm i$ as x becomes arbitrarily large, then we have

$$\frac{\pi}{2} = -\frac{1}{2i}\ln\left(\frac{-i}{i}\right) = \frac{i}{2}[\ln(-i) - \ln(i)] = \frac{i}{2}\left[\ln\left(\frac{1}{i}\right) - \ln(i)\right]$$

$$= \frac{i}{2}[-\ln(i) - \ln(i)] = -i\ln(i)$$

or, once again, $i^i = e^{-\pi/2}$. Once again, like Bernoulli, the Count did not take this last step, but rather concluded his analysis with the amazing expression

$$\frac{\pi}{2} = 2\log\left[(1-\sqrt{-1})^{\frac{1}{2}\sqrt{-1}} \times (1+\sqrt{-1})^{-\frac{1}{2}\sqrt{-1}}\right].$$

But it is easy to show this to be equivalent to $i^i = e^{-\pi/2}$.

If you are perhaps wondering why I've written out all of the tedious, intermediate steps in the evaluation of the imaginary logarithm, it is because there is conceivably an alternative way of doing so which leads to a different answer! That is, why not write

$$\frac{\pi}{2} = -\frac{1}{2i}\ln\left(\frac{-i}{i}\right) = \frac{i}{2}\ln(-1) = \frac{i}{2}\ln(i^2) = i\ln(i) = \ln(i^i)$$

and so $i^i = e^{\pi/2}$? This goes wrong at the step where -1 is replaced with i^2, because that step is ambiguous; we could, just as reasonably, replace -1 with $(-i)^2$. *That* alternative replacement would also lead to the correct result. The way I evaluate the Count's expression, however, is unambiguous at every step, and that's the only way to be sure the final result is correct.

In 1712 Bernoulli used his partial fraction idea to calculate $\tan(n\theta)$ in terms of powers of $\tan(\theta)$, as follows. Define x and y as $x = \tan(\theta)$ and $y = \tan(n\theta)$. Then,

$$\theta = \tan^{-1}(x),$$

$$n\theta = \tan^{-1}(y) = n\tan^{-1}(x).$$

Differentiating the last statement with respect to x,

$$\frac{1}{1+y^2}\frac{dy}{dx} = \frac{n}{1+x^2} \quad \text{or} \quad \frac{dy}{1+y^2} = n\frac{dx}{1+x^2}.$$

Now, using the trick of factoring into imaginary components,

$$\frac{dy}{(y+i)(y-i)} = n\frac{dx}{(x+i)(x-i)},$$

and integrating indefinitely (i.e., no specific limits) the partial fraction expansion of each side, we have

$$\int\frac{1}{2i}\left\{\frac{1}{y-i} - \frac{1}{y+i}\right\}dy = \int\frac{n}{2i}\left\{\frac{1}{x-i} - \frac{1}{x+i}\right\}dx.$$

Writing K as the so-called arbitrary constant of integration (this is absolutely crucial to the rest of the analysis, as you'll soon see),

$$\ln\left(\frac{y-i}{y+i}\right) = n\ln\left(\frac{x-i}{x+i}\right) + K.$$

Since $y = 0$ when $x = 0$ at $\theta = 0$, then

$$\ln\left(\frac{-i}{i}\right) = n\ln\left(\frac{-i}{i}\right) + K.$$

As I stated earlier, it is vital to manipulate such expressions, *always,* in an unambiguous way to solve for K correctly. For example, never replace $-i/i$ with -1 followed by using i^2 for the -1. If you follow this advice you'll find that $K = \ln\{(-1)^{n-1}\}$. That is, K is not "arbitrary" at all. Continuing, we have

$$\ln\left(\frac{y-i}{y+i}\right) = \ln\left(\frac{x-i}{x+i}\right)^n + \ln\{(-1)^{n-1}\} = \ln\left\{(-1)^{n-1}\left(\frac{x-i}{x+i}\right)^n\right\}$$

or

$$\frac{y-i}{y+i} = (-1)^{n-1}\left(\frac{x-i}{x+i}\right)^n.$$

So, since $(-1)^{n-1} = 1$ when n is odd, and $(-1)^{n-1} = -1$ when n is even, we have

$$\frac{y-i}{y+i} = \frac{(x-i)^n}{(x+i)^n}, \quad n \text{ odd } (1, 3, 5, \ldots),$$

$$\frac{y-i}{y+i} = \frac{(x-i)^n}{(x+i)^n}, \quad n \text{ even } (2, 4, 6, \ldots).$$

Solving for y gives us what we want:

$$y = i\frac{(x+i)^n + (x-i)^n}{(x+i)^n - (x-i)^n}, \quad n \text{ odd,}$$

$$y = i\frac{(x+i)^n - (x-i)^n}{(x+i)^n + (x-i)^n}, \quad n \text{ even.}$$

These expressions can be expanded using the binomial theorem to remove all the i's, but I'll leave that as an exercise for you. For any given "small" value of n, the expressions are easy to evaluate directly. For example, for $n = 4$ and $n = 5$, respectively, these formulas give

$$tan(4\theta) = \frac{4\tan(\theta) - 4\tan^3(\theta)}{1 - 6\tan^2(\theta) + \tan^4(\theta)},$$

$$tan(5\theta) = \frac{5\tan(\theta) - 10\tan^3(\theta) + \tan^5(\theta)}{1 - 10\tan^2(\theta) + 5\tan^4(\theta)},$$

expressions that can be found in any comprehensive mathematics handbook.

Care with the constant of integration K is, as I have emphasized, of central importance in this analysis. Even Bernoulli himself could fall into this trap. For example, in a correspondence during 1712–13 with Leibniz, Bernoulli claimed that since $dx/x = -dx/-x = d(-x)/(-x)$, then integration gives $\log(x) = \log(-x)$. Thus, since $\log(1) = 0$ then it follows that $\log(-1) = 0$. And since $\log(-1) = \log\{(\sqrt{-1})^2\} = 2\log(\sqrt{-1})$, then $\log(\sqrt{-1}) = 0$, too. Bernoulli's error here is, of course, as you must now suspect, that he has forgotten the constant of integration. Leibniz felt Bernoulli was wrong, but for the wrong reason. He argued that $\log(-1)$ could not be zero because if it were then $\log(\sqrt{-1})$ would be half of that, which is right, but certainly (said Leibniz) $\log(\sqrt{-1})$ does not have *any* value—he called $\log(\sqrt{-1})$ "imaginary," which is right in today's terminology, but Leibniz actually meant "nonexistent" by that word.

6.8 ROGER COTES AND A LOST OPPORTUNITY

Of all of Newton's English contemporaries, one of the least well known to modern engineers and scientists is Roger Cotes (1682–1716), whose death from a violent fever one month before his thirty-fourth birthday cut short what was truly a promising life. You'll recall we encountered him, or at least one of his theorems, back in chapter 4. A professor at Cambridge by age twenty-six, he was also editor of the second edition of Newton's masterpiece *Principia,* a work that revolutionized physics. It was reported that, after his death, Newton himself said of Cotes "If he had lived we might have known something."

Actually, Newton was wrong. *Before* he died Cotes had already published, in 1714, a result that told the world something of great significance, and which would have earned him immortal fame—if only he had written a bit more clearly. This result, also discovered by another of Newton's friends, Abraham De Moivre, who, it seems, knew the result years before, and then later by Euler, is nothing less than what I called Euler's identity at the start of this chapter. With just a little more attention by Cotes to his prose, however, it could easily be known today as Cotes' identity, and he would be a celebrated saint to all electrical engineers, physicists, and mathematicians. As it is, most of them have heard of him perhaps only once or twice, and then have promptly forgotten just why his name came up in the first place.

As were nearly all mathematicians of his day, Cotes was a geometer by nature and his derivations are shot through with drawings of lines, circles, and other more complicated curves, and much talk of geometric construction. What I'll show you here, however, is a modern, analytical presentation of how I have reconstructed what Cotes did. Imagine, then, an ellipse given by the usual formula

$$\frac{x^2}{a^2} + \frac{y^2}{b^2} = 1$$

where a and b are the lengths of the semi-axes. If you take just that part of the ellipse in the first quadrant and revolve it around the y-axis, then you will generate the surface of the top half of an ellipsoid. What Cotes did was to derive a formula for the surface area of this "surface of revolution." Indeed, he derived two such formulas. From this rather ordinary beginning comes a singularly astonishing result.

The surface area generated is, with reference to the notation of figure 6.1,

$$A = \int 2\pi x \ ds, \ ds = \sqrt{(dx)^2 + (dy)^2},$$

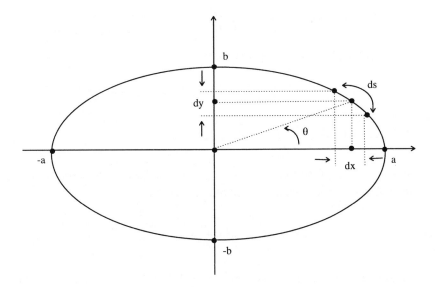

Figure 6.1. Cotes' ellipse.

where ds is the differential of arc length along the ellipse being revolved about the y-axis. Letting $x = a\cos(\theta)$ and $y = b\sin(\theta)$, we have $dx = -a\sin(\theta)\,d\theta$ and $dy = b\cos(\theta)\,d\theta$, and so $ds = \sqrt{a^2\sin^2(\theta) + b^2\cos^2(\theta)}\,d\theta$. Thus,

$$a = \int_0^{\frac{\pi}{2}} 2\pi a\cos(\theta)\sqrt{a^2\sin^2(\theta) + b^2\cos^2(\theta)}\ d\theta.$$

Now, let $u = \sin(\theta)$ and so $du = \cos(\theta)\,d\theta$. Then,

$$A = \int_0^1 2\pi a\sqrt{1-u^2}\ \sqrt{a^2 u^2 + b^2(1-u)^2}\ \frac{du}{\sqrt{1-u^2}}$$

$$= 2\pi a \int_0^1 \sqrt{u^2(a^2 - b^2) + b^2}\ du.$$

Notice, carefully, that up to this point nothing has been said about the relative magnitudes of a and b. I drew figure 6.1 with $a > b$, but I could just as well have drawn it with $a < b$. There are, in fact, two choices open to us here. So, if we wish to express A in terms of real integrals, here is what we must write:

$$A = 2\pi a\sqrt{a^2 - b^2} \int_0^1 \sqrt{u^2 + \frac{b^2}{a^2 - b^2}}\ du \text{ if } a > b,$$

$$A = 2\pi a\sqrt{b^2 - a^2} \int_0^1 \sqrt{\frac{b^2}{b^2 - a^2} - u^2}\ du \text{ if } a < b.$$

From integral tables you can verify that

$$\int \sqrt{x^2 + c^2}\ dx = \frac{x\sqrt{x^2 + c^2}}{2} + \frac{c^2}{2}\ln\left(x + \sqrt{x^2 + c^2}\right),$$

$$\int \sqrt{c^2 - x^2}\ dx = \frac{x\sqrt{c^2 - x^2}}{2} + \frac{c^2}{2}\sin^{-1}\left(\frac{x}{c}\right).$$

Using the first of these formulas on the $a > b$ case [with $c^2 = b^2/(a^2 - b^2)$, and applying a little algebra, it is routine to show that

$$A = \pi a\left\{a + \frac{b^2}{\sqrt{a^2 - b^2}}\ln\left(\frac{a + \sqrt{a^2 - b^2}}{b}\right)\right\} \text{ if } a > b.$$

And using the second of the integration formulas on the $a < b$ case [now with $c^2 = b^2/(b^2 - a^2)$, we arrive at

$$A = \pi a \left\{ a + \frac{b^2}{\sqrt{b^2 - a^2}} \sin^{-1}\left(\frac{\sqrt{b^2 - a^2}}{b} \right) \right\} \quad \text{if } a < b.$$

It is important to understand why I have written two expressions for A—it is purely for the reason of having only real quantities in each equation. But that is simply an arbitrary motivation, as either expression is correct for both $a > b$ and $a < b$. If one does get imaginaries at one point in evaluating one of the equations for A, then there will be another imaginary somewhere else that will "eliminate" the first one. A, a physical area, *must* be real.

Now, continuing on, let's define the quantity ϕ such that

$$\sin(\phi) = \frac{\sqrt{b^2 - a^2}}{b} = \frac{i\sqrt{a^2 - b^2}}{b}.$$

In either case, $\cos(\phi) = a/b$. With this we can now write our two expressions for A as

$$A = \pi a \left\{ a + \frac{b^2}{\sqrt{a^2 - b^2}} \ln\left(\frac{a}{b} + \frac{\sqrt{a^2 - b^2}}{b} \right) \right\}$$

$$= \pi a \left\{ a + \frac{b^2}{\sqrt{a^2 - b^2}} \ln[\cos(\phi) - i\sin(\phi)] \right\}, \quad a > b,$$

and

$$A = \pi a \left\{ a + \frac{b^2}{i\sqrt{a^2 - b^2}} \sin^{-1}[\sin(\phi)] \right\} = \pi a \left\{ a + \frac{b^2}{\sqrt{a^2 - b^2}} (-i\phi) \right\}, \quad a < b.$$

Comparing these two expressions, *which represent the same physical quantity,* it is immediately obvious that $-i\phi = \ln[\cos(\phi) - i\sin(\phi)]$, or

$$e^{-i\phi} = \cos(\phi) - i\sin(\phi),$$

or, equivalently,

$$e^{i\phi} = \cos(\phi) + i\sin(\phi).$$

Thus could Cotes have discovered Euler's identity.

But Cotes did *not* take this last step and his $-i\phi = \ln[\cos(\phi) - i\sin(\phi)]$ simply appears embedded, in almost unbelievably obscure language, in the only paper he wrote for publication in his entire life ("Logometria," which appeared in the March 1714 issue of the *Philosophical Transactions* of the Royal Society).[4] This paper is notable, too, for Cotes' calculation of both e

and $1/e$ (each correct to twelve decimal places) using power series expansions. This was in 1714, remember, decades before Euler calculated e to twenty-three places in 1748. Cotes was clearly an enormously gifted intellect, but still, after reading Cotes I think you'll agree that to say he was "obscure" is generous.

Even when "Logometria" was combined with other of Cotes' work and published posthumously as the book *Harmonia Mensurarum* (1722), its brilliant contents apparently went unnoticed, or at least the ellipsoidal calculation certainly did. After a final comment about multiplying a quantity by $\sqrt{-1}$, which certainly must have puzzled some readers in 1714, Cotes breaks off his derivation with the odd words "but I leave this to be examined in more detail by others who think it worthwhile." For nearly two centuries after, nobody thought it was. Indeed, it went unnoticed for 185 years, until a Russian mathematician brought it to the world's attention in an 1899 book on the history of functions. The reasons for such a monumental oversight are not completely clear, but Cotes was well known for giving the answers to problems while providing almost no explanation for his readers about where they came from. His readers at the time probably simply didn't understand what Cotes meant. Let this be a lesson in the value of clear exposition!

6.9 Many-Valued Functions

At this point I should tell you that while it is now clear that Euler was not the first to consider i^i, it *was* Euler who first observed that i^i is real and has an infinity of values. That is, $e^{-\pi/2}$ is just one possible value. This is because of the geometric fact that, starting with any angle, adding or subtracting any multiple of 2π radians to it simply swings you through an integer number of complete rotations about the origin and you end up back where you started. We saw the same sort of thing back in box 3.2. Thus, with n any integer (positive, negative, or zero),

$$i^i = \left\{ e^{i\left(\frac{1}{2}\pi + 2\pi n\right)} \right\}^i = e^{-\left(\frac{1}{2}\pi + 2\pi n\right)} = e^{-\left(\frac{1}{2} + 2n\right)\pi}.$$

It is customary to call the $n = 0$ case the *principal value* of i^i, but it isn't the only value. For $n = -1$, 1, and 2, for example, we get, respectively, 111.3178, 3.882×10^{-4}, and 7.2495×10^{-7}.

Even more astonishing, perhaps, is that 1^π—a real number to a real power—has an infinity of distinct *complex* values, i.e.,

$$1^\pi = \cos(2\pi^2 n) + i\,\sin(2\pi^2 n),\ n = 0,\ \pm1,\ \pm2,\ \ldots\ .$$

Only for $n = 0$, the principal value, is 1^π real. This surprising result is due to π being irrational, a result (1761) due to the German mathematician Johann Lambert (1728–71). This is because $2\pi^2 n = \pi(2\pi n)$, and for $n \neq 0$ $2\pi n$ can never be an integer—if it could be then π would be rational. Since, except for $n = 0$, $2\pi n$ can never be an integer, then $\sin(2\pi^2 n)$ can never vanish, i.e., 1^π will always have a nonzero imaginary part (except for $n = 0$, of course). Further, using the same general argument, it is easy to show that there are no distinct integer values for n that result in either the real or imaginary parts of 1^π repeating. Again, if there were, π would be rational. Finally, since Euler showed (1737) that e is irrational, then 1^e has an infinity of distinct complex values, too, as does $1^{\sqrt{n}}$ where n is any positive, nonsquare integer greater than one (recall Theaetetus' irrationality result from section 2.1). It is important to understand that the complex nature of $1^{\text{irrational}}$ is a theoretical result, a result that no actual computer made of a finite amount of matter could ever exhibit. In any physically constructable machine, using a finite number of digits to represent numbers, all numbers are necessarily rational.

In the same way as with $1^{\text{irrational}}$, it is now easy to see that the logarithm is a multivalued function, too. Thus, write the arbitrary complex number $a + ib$ in polar form as

$$a + ib = \sqrt{a^2 + b^2}\ \angle\ \tan^{-1}\!\left(\frac{b}{a}\right).$$

We have, with n any integer,

$$a + ib = \sqrt{a^2 + b^2}\ e^{i\{\tan^{-1}(b/a) + 2\pi n\}}.$$

Thus,

$$\ln(a + ib) = \frac{1}{2}\ln(a^2 + b^2) + i\left\{\tan^{-1}\!\left(\frac{b}{a}\right) + 2\pi n\right\},\ n = 0,\ \pm1,\ \pm2,\ \ldots\ .$$

Also, as before, the $n = 0$ case gives the so-called principal value of the log function. So, for example, the principal value of $\ln(1 + i) = \frac{1}{2}\ln(2) + i\pi/4 = 0.346573 + i0.785398$.

6.10 THE HYPERBOLIC FUNCTIONS

With the discovery by Euler of the exponential expressions for the sine and cosine trigonometric functions, sense could then be made of complex angles, even though we can not actually draw, or even visualize, such angles. That is, if we simply plunge right into the symbols, we have

$$\cos(x+iy) = \frac{e^{i(x+iy)} + e^{-i(x+iy)}}{2} = \frac{e^{ix}e^{-y} + e^{-ix}e^{y}}{2}$$

$$= \frac{e^{-y}\{\cos(x)+i\sin(x)\}+e^{y}\{\cos(x)-i\sin(x)\}}{2}$$

$$= \cos(x)\left\{\frac{e^{y}+e^{-y}}{2}\right\} - i\sin(x)\left\{\frac{e^{y}-e^{-y}}{2}\right\}$$

$$= \cos(x)\cosh(y) - i\sin(x)\sinh(y),$$

where the *hyperbolic cosine* and *hyperbolic sine* are defined to be, respectively,

$$\cosh(y) = \frac{1}{2}\left\{e^{y}+e^{-y}\right\},$$

$$\sinh(y) = \frac{1}{2}\left\{e^{y}-e^{-y}\right\}.$$

In the same way, you can easily verify that

$$\sin(x+iy) = \sin(x)\cosh(y) + i\cos(x)\sinh(y).$$

Notice that if $x = 0$ then $\cos(iy) = \cosh(y)$ and $\sin(iy) = i\sinh(y)$, i.e., the cosine of an imaginary angle is *real*, while the sine of an imaginary angle is also imaginary. Finally, in analogy to the ordinary trigonometric identity of $\tan(\theta) = \sin(\theta)/\cos(\theta)$, we can define the *hyperbolic tangent* as

$$\tanh(x+iy) = \frac{\sinh(x+iy)}{\cosh(x+iy)}.$$

The hyperbolic functions are so useful in science and engineering that they have been tabulated in mathematical handbooks, and are available as special key functions on any scientific electronic pocket calculator costing more than fifteen dollars. They were introduced into general use by the Italian mathematician Vincenzo Riccati (1707–75), in his two-volume work *Opusculorum ad Res Physicas et Mathematicas Pentinentium* (1757–62). The term "hyperbolic" comes from the fact that if we define the Cartesian coordinates of a point to be $x = \cosh(t)$ and $y = \sinh(t)$, where t is a so-called *parametric variable* (generally, $-\infty < t < \infty$), then elimination of t leads to the equation $x^2 - y^2 = 1$, the equation of an hyperbola. The "sine" and "cosine" come both from the suggestive similarity of the hyperbolic functions' exponential definitions to those of the sine and cosine of the circular trigonometric functions, and from the superficial similarity of the circle's equation, $x^2 + y^2 = 1$, to that of the hyperbola.

One of the virtues of the hyperbolic functions is that they allow us to remove the usual restrictions of $|\sin(\theta)| \leq 1$ and $|\cos(\theta)| \leq 1$, restrictions you may recall made back in Section 1.6 when discussing Viète's trigonometric solution for cubic equations. These restrictions hold only if θ is real—if θ is allowed to be complex (whatever that might mean), well then, things get interesting! As an example, let us calculate the angle that has a cosine of 2, i.e., the complex value of θ such that $\cos(\theta) = 2$. Looking back at the expression for $\cos(x + iy)$, we can set its real part equal to 2 and the imaginary part to zero, i.e.,

$$\sin(x)\sinh(y) = 0,$$
$$\cos(x)\cosh(y) = 2.$$

These two conditions are satisfied by picking $x = 2\pi n$ (with n any integer) and $\cosh(y) = 2$. This leads immediately to the equation

$$e^{2y} - 4e^y + 1 = 0,$$

which is a quadratic in e^y. Thus, $y = \ln(2 \pm \sqrt{3})$. The answer to our original question, then, is

$$\theta = \cos^{-1}(2) = 2\pi n + i \ln(2 \pm \sqrt{3}).$$

A natural feeling at this point is that we are perhaps playing some kind of symbol-pushing game here, simply writing down equations without regard for whether or not they make sense. One way to demonstrate the usefulness of hyperbolic functions is to actually calculate something we can verify by other means. So, to that end, let me illustrate what I am talking about by first showing you a related calculation: suppose you are asked to calculate the value of

$$S = \sum_{n=1}^{\infty} \tan^{-1}\left(\frac{2}{n^2}\right).$$

This particular problem has a long history, dating from its first appearance in the scholarly literature in 1878. By the early 1900s the Indian genius Srinivasa Ramanujan (1887–1920) took enough interest in it to work out the value of S. We've even seen something similar to it in chapter 2, in connection with Theodorus' spiral of triangles. Today we can write a simple computer program to evaluate S in a flash (the computer answer is 2.356), but is there an analytical solution? There is indeed.

Define the two angles α and β to be such that $\tan(\alpha) = n + 1$ and $\tan(\beta) = n - 1$. Then, using the trigonometric identity for $\tan(\alpha - \beta)$, we have

169

$$\tan(\alpha - \beta) = \frac{\tan(\alpha) - \tan(\beta)}{1 + \tan(\alpha)\tan(\beta)} = \frac{(n+1) - (n-1)}{1 + (n+1)(n-1)} = \frac{2}{n^2}.$$

Thus,

$$\alpha - \beta = \tan^{-1}(n+1) - \tan^{-1}(n-1) = \tan^{-1}\left(\frac{2}{n^2}\right)$$

and so

$$S = \sum_{n=1}^{\infty} \{\tan^{-1}(n+1) - \tan^{-1}(n-1)\}.$$

Writing this out term by term, you can easily verify that the Nth partial sum is

$$S_N = \tan^{-1}(N + 1) + \tan^{-1}(N) - \tan^{-1}(1) - \tan^{-1}(0).$$

So

$$S = \lim_{N \to \infty} S_N = \frac{1}{2}\pi + \frac{1}{2}\pi - \frac{1}{4}\pi = \frac{3}{4}\pi = 2.35619\ldots,$$

which agrees with the computer answer. The fact that we are able to calculate S using the tangent and inverse tangent functions and get a numerical result that agrees with direct computer computation certainly enhances one's "feelings of comfort" about those particular trigonometric functions. Wouldn't it be nice to be able to do something similar with the hyperbolics?

Suppose, then, that a seemingly slight change is made in the problem. Now you are to calculate

$$T = \sum_{n=1}^{\infty} \tan^{-1}\left(\frac{1}{n^2}\right).$$

Again, this is duck soup for a computer, and simply changing the program code from $2/n^2$ to $1/n^2$ leads to the computed answer of $T = 1.4245$. But now the little algebraic trick I used to express $\tan^{-1}(2/n^2)$ in terms of $\tan^{-1}(n + 1)$ and $\tan^{-1}(n - 1)$ doesn't work. What to do?

Visualizing a right triangle with a base of 1 and a height of $1/n^2$ allows us to write $\tan^{-1}(1/n^2)$ as the polar angle of the complex number $1 + i1/n^2$. Thus,

$$T = \sum_{n=1}^{\infty} \tan^{-1}\left(\frac{1}{n^2}\right) = \sum_{n=1}^{\infty} \angle\left(1 + i\frac{1}{n^2}\right).$$

While it may not be instantly obvious *why* the following will be useful, it certainly is true that, as you can quickly verify,

$$1 + i\frac{1}{n^2} = 1 + \frac{\{\pi(1+i)/\sqrt{2}\}^2}{\pi^2 n^2}.$$

Now, since the angle of a product of complex numbers is the sum of the individual angles of each complex number factor, then

$$T = \sum_{n=1}^{\infty} \angle\left(1 + i\frac{1}{n^2}\right) = \angle \prod_{n=1}^{\infty}\left[1 + \frac{\{\pi(1+i)/\sqrt{2}\}^2}{\pi^2 n^2}\right].$$

This last expression may look familiar to you—it is of the same form as Euler's product for sin(y) if we write

$$y^2 = -\{\pi(1 + i)/\sqrt{2}\}^2.$$

Thus,

$$T = \angle\frac{\sin(y)}{y} = \angle\frac{\sin[i\{\pi(1+i)/\sqrt{2}\}]}{i\{\pi(1+i)/\sqrt{2}\}}.$$

The numerator of the last expression is the sine of a complex argument, and so we expect hyperbolics to appear. In fact, since $\sin(x + iy) = \sin(x)\cosh(y) + i\cos(x)\sinh(y)$, then

$$T = \angle\frac{\sin\left(-\dfrac{\pi}{\sqrt{2}} + i\dfrac{\pi}{\sqrt{2}}\right)}{-\dfrac{\pi}{\sqrt{2}} + i\dfrac{\pi}{\sqrt{2}}} = \angle\frac{\left\{\sin\left(-\dfrac{\pi}{\sqrt{2}}\right)\cosh\left(\dfrac{\pi}{\sqrt{2}}\right) + i\cos\left(\dfrac{\pi}{\sqrt{2}}\right)\sinh\left(\dfrac{\pi}{\sqrt{2}}\right)\right\}}{-\dfrac{\pi}{\sqrt{2}} + i\dfrac{\pi}{\sqrt{2}}}.$$

The angle of the ratio of two complex numbers is, of course, the difference between the angles of the numerator and the denominator. The denominator angle is clearly $\frac{3}{4}\pi$ since the denominator is a complex number in the second quadrant. And a careful evaluation of the numerator shows it to be a complex number in the third quadrant. That is,

$$T = \angle\{-3.711536874 - i\,2.759617006\} - \frac{3}{4}\pi$$

$$= \pi + \tan^{-1}\left\{\frac{2.759617006}{3.711536874}\right\} - \frac{3}{4}\pi$$

$$= \frac{1}{4}\pi + 0.639343615 \cdots = 1.424741778 \cdots.$$

This calculated value of T is very close to the program-computed value of T, and we got it with hyperbolics and the sine of a complex angle.

Finally, we can write the hyperbolics as infinite products by substituting iy for y, e.g., from section 6.5 we have

$$\sin(iy) = i\sinh(y) = iy \prod_{n=1}^{\infty}\left(1 + \frac{y^2}{\pi^2 n^2}\right)$$

or

$$\sinh(y) = y \prod_{n=1}^{\infty}\left(1 + \frac{y^2}{\pi^2 n^2}\right).$$

To test this on a programmable calculator, I compared $\sinh(2) = 3.62686$ with the result of using the first 10, and then the first 1,000 factors, of the product: the results were, respectively, 3.48972 and 3.62537. The convergence is clear, but it *is* slow.

6.11 CALCULATING π FROM $\sqrt{-1}$

Amazingly, the quite formal and "mysterious formula" of $\pi = (2/i)\ln(i)$, as Benjamin Peirce called it, can be used to calculate the numerical value of π. That might seem like getting something out of thin air, but this astonishing fact was pointed out long ago by the German mathematician and educator Karl Heinrich Schellbach (1809–90) in 1832. Schellbach wrote

$$\frac{\pi i}{2} = \ln(i) = \ln\left(\frac{1+i}{1-i}\right) = \ln(1+i) - \ln(1-i)$$

and then expanded the two logarithms into their power series. That is, even though i is not real he simply substituted $z = i$ into Mercator's formula from section 6.3

$$\ln(1+z) = z - \frac{1}{2}z^2 + \frac{1}{3}z^3 - \frac{1}{4}z^4 + \cdots,$$

which gives

$$\ln(1+i) = i + \frac{1}{2} - \frac{1}{3}i - \frac{1}{4} + \frac{1}{5}i + \frac{1}{6} - \cdots.$$

Similarly,

$$\ln(1-i) = -i + \frac{1}{2} + \frac{1}{3}i - \frac{1}{4} - \frac{1}{5}i + \frac{1}{6} + \cdots.$$

Thus,

$$\frac{\pi i}{2} = 2i - \frac{2}{3}i + \frac{2}{5}i - \cdots$$

or

$$\frac{\pi}{4} = 1 - \frac{1}{3} + \frac{1}{5} - \frac{1}{7} + \frac{1}{9} - \frac{1}{11} + \cdots$$

a result discovered by an entirely different method in 1674 by Leibniz, after whom the series is usually named, although in fact it was known even earlier—the Scottish mathematician James Gregory (1638–75), for example, discovered it for himself in 1671.

The Leibniz-Gregory series is, while beautifully elegant in appearance, utterly worthless for numerical calculations since it converges very slowly. Using the first fifty-three terms, for example, is not sufficient to give even just two correct, stable decimal digits. When running a program that continually displays the partial sums on a monitor screen while summing the terms of a convergent series, you will of course see all the digits flicker wildly at first. But, gradually, the digits will settle down from left to right and stop changing no matter how many more terms of the series are subsequently included. When a digit stops changing, it is *stable*. Don't confuse the *error* in the value of a partial sum with the number of correct digits—the two are quite different things. A result in the theory of convergent series, such as the Leibniz-Gregory series, says that a partial sum is correct within an error less than the first term neglected. That is, if we keep the first five terms of the Leibniz-Gregory series then the partial sum is 0.8349206, and that is within $\frac{1}{11}$—which is less than 0.1—of the true value of $\pi/4 = 0.7853982. \ldots$ But notice that *not a single digit* of the partial sum, not even the first, is correct.

Now, what Schellbach went on to show was how his method gives other series for π that converge much faster than does the Leibniz-Gregory series. For example, he wrote

$$\frac{\pi i}{2} = \ln(i) = \ln\left\{\frac{(2+i)(3+i)}{(2-i)(3-i)}\right\} = \ln\left\{\frac{\left(1+\frac{1}{2}i\right)\left(1+\frac{1}{3}i\right)}{\left(1-\frac{1}{2}i\right)\left(1-\frac{1}{3}i\right)}\right\}$$

$$= \left\{\ln\left(1+\frac{1}{2}i\right) - \ln\left(1-\frac{1}{2}i\right)\right\} + \left\{\ln\left(1+\frac{1}{3}i\right) - \ln\left(1-\frac{1}{3}i\right)\right\}$$

and then, as before, expanded the logarithms. This gives

$$\frac{\pi}{4} = \left(\frac{1}{2}+\frac{1}{3}\right) - \frac{1}{3}\left(\frac{1}{2^3}+\frac{1}{3^3}\right) + \frac{1}{5}\left(\frac{1}{2^5}+\frac{1}{3^5}\right) - \frac{1}{7}\left(\frac{1}{2^7}+\frac{1}{3^7}\right) + \cdots,$$

which clearly converges faster than does the Leibniz-Gregory series. Indeed, notice that this new series *is* the Leibniz-Gregory series, but with the terms multiplied by factors less than one which are, themselves, rapidly approaching zero. If we keep the first five terms of this series the partial sum is 0.7854353 and we have three correct, stable digits. If you keep just four more terms (0.7853983) then you'll have six correct, stable digits.

There is no need to stop with this series, however. One can continue on, indefinitely, using even more complicated expressions for the i in the $\ln(i)$ in the original "mysterious" formula. Schellbach suggested, for example, the two expressions

$$\pi = \frac{2}{i}\ln\left\{\frac{(5+i)^4(-239+i)}{(5-i)^4(-239-i)}\right\} = \frac{2}{i}\ln\left\{\frac{(10+i)^3(-515+i)^4(-239+i)}{(10-i)^3(-515-i)^4(-239-i)}\right\}.$$

If the first expression is expanded in the way I've done earlier, then you can show that

$$\frac{\pi}{4} = 4\left[\frac{1}{5} - \frac{1}{3\cdot5^3} + \frac{1}{5\cdot5^5} - \frac{1}{7\cdot5^7} + \cdots\right] - \left[\frac{1}{239} - \frac{1}{3\cdot239^3} + \frac{1}{5\cdot239^5} - \cdots\right],$$

a result discovered in 1706, using entirely different means, by the London astronomer John Machin (1680–1752), who used it to calculate π to one hundred decimal places. This result is actually only a slightly disguised form of a result from section 3.1, where I asked you to use Wessel's "add the angles" rule in multiplying complex numbers, to confirm the formula

$$\frac{\pi}{4} = 4\tan^{-1}\left(\frac{1}{5}\right) - \tan^{-1}\left(\frac{1}{239}\right).$$

The Schellbach/Machin formula is simply this result with the inverse tangent functions expanded as power series. Nearly 250 years after Machin, the world's first electronic computer (ENIAC, for Electronic Numerical Integrator And Calculator) used the same formula to calculate π to over two thousand decimal places. What a *miracle* this all is, with the ever-changing decimal digits for π flowing endlessly from the square root of minus one.

6.12 USING THE COMPLEX TO DO THE REAL

As dramatic illustrations of the utility of complex numbers *before* Wessel, let me give you some more examples of the genius of Euler. Imagine it is 1743 and you are a mathematician at the peak of your powers. You are confronted by the following integrals, integrals never before encountered by mathematicians:

$$I_1 = \int_0^\infty \sin(s^2)ds \quad \text{and} \quad I_2 = \int_0^\infty \cos(s^2)ds.$$

How would you proceed? I don't think it's at all obvious just what to do.

While primarily a mathematician, Euler was often stimulated in his mathematics by questions raised during the study of physical problems. In this case, the above two integrals occurred in an analysis of the physics of a coiled spring.[5] Much later, in 1815, the French mathematicians Cauchy and Poisson developed equivalent integrals in certain hydrodynamical applications.[6] Three years later, in 1818, the French scientist Augustin Jean Fresnel (1788–1827) also ran up against I_1 and I_2 in his study of the diffraction of light, and today they are usually called the Fresnel integrals. One does still see them also called the Euler integrals, however, and it was Euler who first evaluated them. But it wasn't easy to do these evaluations, even for a man with Euler's powerful intellect. As he wrote of them in 1743, "We must admit that analysis will make no small gain should anyone find a method whereby, approximately at least, the value of [these integrals] would be determined. . . . This problem does not seem to be unworthy of the best strength of geometers."

The best that even Euler could do in 1743, however, was to find convergent infinite series for I_1 and I_2 that allowed him to make numerical calculations; nearly forty years later he solved the problem exactly using complex quantities. What I will show you next is essentially what Euler reported to the St. Petersburg Academy on April 30, 1781, but which was published only years after his death. The key idea to evaluating I_1 and I_2 is to start with yet another of Euler's creations, with what today is called the *gamma function*.[7] It is defined as

$$\Gamma(n) = \int_0^\infty e^{-x}x^{n-1}dx, \quad n > 0.$$

That is, n is restricted to be any positive number. It is clear, for $n = 1$, by direct integration, that

$$\Gamma(1) = \int_0^\infty e^{-x}dx = 1.$$

It is also not hard to show, using integration by parts, the so-called functional formula

$$\Gamma(n + 1) = n\Gamma(n).$$

So, for n a positive integer, we have a connection between $\Gamma(n)$ and the factorial function:

$$\Gamma(2) = 1 \cdot \Gamma(1) = 1 \cdot 1 = 1!$$
$$\Gamma(3) = 2 \cdot \Gamma(2) = 2 \cdot 1! = 2!$$
$$\Gamma(4) = 3 \cdot \Gamma(3) = 3 \cdot 2! = 3!$$

.

.

.

$$\Gamma(n) = (n-1) \cdot (n-2)! = (n-1)!$$

Setting $n = 1$, the last statement says $\Gamma(1) = 0!$, but since we have already shown $\Gamma(1) = 1$ then we have $0! = 1$, as claimed back in section 3.2.

We can extend the result $\Gamma(n) = (n-1)!$ to all real n, including negative n, by using the recursion relation $\Gamma(n) = (1/n)\,\Gamma(n+1)$ to work backwards into the negative values of n. For example,

$$\Gamma\left(-\frac{1}{2}\right) = \frac{1}{-\frac{1}{2}}\Gamma\left(\frac{1}{2}\right) = -2\sqrt{\pi},$$

because, as I'll show you next, $\Gamma(\frac{1}{2}) = \sqrt{\pi}$. It is curious to note that a closely related result was anticipated by John Wallis long before Euler. The result $(\frac{1}{2})! = \frac{1}{2}\sqrt{\pi}$, which follows from $\Gamma(\frac{3}{2}) = (\frac{1}{2})! = \frac{1}{2}\Gamma(\frac{1}{2}) = \frac{1}{2}\sqrt{\pi}$, was known to Wallis even though he, of course, knew nothing of the gamma integral. Here's how he knew this. After doing some specific evaluations of the integral $\int_0^1 (x - x^2)^n dx$ for particular integer values of n, Wallis determined that the general value of this integral is $(n!)^2/(2n+1)!$ He also knew the value for the fractional case of $n = \frac{1}{2}$, i.e., $\int_0^1 (x - x^2)^{1/2}dx = \pi/8$, because this integral physically represents the area enclosed by the top half of the circle centered on the x-axis at $x = \frac{1}{2}$ with a diameter of one, and the x-axis. Then, making the giant assumption that the general formula for integer n holds for fractional n as well, Wallis wrote $\{(\frac{1}{2})!\}^2/2! = \pi/8$, from which it follows that $(\frac{1}{2})! = \frac{1}{2}\sqrt{\pi}$.

In Euler's integral definition for $\Gamma(n)$, of course, n does *not* have to be restricted to integer values. Indeed, it was the question of how to interpolate the factorial function (e.g., what is $(5\frac{1}{2})!?$) that motivated Euler to develop the integral definition in the first place. The integral definition allows the calculation of noninteger factorials. For example, we can write something as mysterious looking as $(-\frac{1}{2})!$ and actually have it mean something, i.e., for $n = \frac{1}{2}$ we have

$$\Gamma\left(\frac{1}{2}\right) = \left(-\frac{1}{2}\right)! = \int\limits_{0}^{\infty} \frac{e^{-x}}{\sqrt{x}} dx.$$

Now, this might look as though we have replaced one mystery [what is $(-\frac{1}{2})!?$] with another (what is the integral?), but this integral *can* be evaluated. There is a pretty little trick to the calculation, and it is worth examining the details because I'll actually use the value of $\Gamma(\frac{1}{2})$ in completing Euler's evaluation of the Fresnel integrals via complex numbers.

Begin by making the change of variable $x = t^2$. Then $dx = 2t\,dt$ and so

$$\Gamma\left(\frac{1}{2}\right) = \int\limits_{0}^{\infty} \frac{e^{-x}}{\sqrt{x}} dx = \int\limits_{0}^{\infty} \frac{e^{-t^2}}{t} 2t\,dt = 2\int\limits_{0}^{\infty} e^{-t^2} dt = 2I,$$

where

$$I = \int\limits_{0}^{\infty} e^{-u^2} du = \int\limits_{0}^{\infty} e^{-v^2} dv.$$

Of course, the only difference between the various integral expressions for I is the trivial one of the particular symbol—t, u, or v—used for the dummy variable of integration. Now, write

$$I^2 = \left\{\int\limits_{0}^{\infty} e^{-u^2} du\right\}\left\{\int\limits_{0}^{\infty} e^{-v^2} dv\right\} = \int\limits_{0}^{\infty}\int\limits_{0}^{\infty} e^{-(u^2+v^2)} du\ dv.$$

This does look pretty horrible, but there's a nifty trick that every mathematician, engineer, and physicist should know that melts away all the apparent difficulty. The double integral is, geometrically, the integration of a function $f(u,v)$ over the first quadrant of the u,v-plane, i.e., over $0 \le u < \infty, 0 \le v < \infty$, with $du\,dv$ as the differential area in Cartesian coordinates. We will obviously change nothing *physically* by a change in coordinates. If we change to polar coordinates, in particular, then the mathematics simplifies tremendously. So, let us write $u = r\cos(\theta)$ and $v = r\sin(\theta)$, with the corresponding differential area now $r\,dr\,d\theta$. To cover the first quadrant we require $0 \le r < \infty$ and $0 \le \theta < \pi/2$. Since $u^2 + v^2 = r^2$, then

$$I^2 = \int\limits_{0}^{\frac{\pi}{2}}\int\limits_{0}^{\infty} e^{-r^2} r\ dr\ d\theta.$$

The interior r-integral is elementary, i.e.,

$$\int\limits_{0}^{\infty} e^{-r^2} r\ dr = -\frac{1}{2} e^{-r^2}\Big|_{0}^{\infty} = \frac{1}{2},$$

and so

$$I^2 = \int_0^{\frac{\pi}{2}} \frac{1}{2} \, d\theta = \frac{\pi}{4}.$$

That is, $I = \frac{1}{2}\sqrt{\pi}$ and as $\Gamma(\frac{1}{2}) = 2I$ then

$$\Gamma\left(\frac{1}{2}\right) = \int_0^\infty \frac{e^{-x}}{\sqrt{x}} \, dx = \sqrt{\pi} = \left(-\frac{1}{2}\right)!.$$

Looking back at the start of this calculation we see that another way of writing this result is as

$$I = \int_0^\infty e^{-u^2} \, du = \frac{1}{2}\sqrt{\pi},$$

or as

$$\int_{-\infty}^\infty e^{-x^2} \, dx = \sqrt{\pi}.$$

This result was known to De Moivre since at least 1733, during the course of his studies in the theory of probability. There is a marvelous story that once, when giving a lecture to engineering students, Lord Kelvin used the word "mathematician" and then stopped, looked at his class, and asked "Do you know what a mathematician is?" He then wrote the above integral on the blackboard and said "A mathematician is one to whom *that* is as obvious as that twice two makes four is to you." I think Kelvin was reaching if he really said that, but still, if he didn't he should have because it *is* a great story!

Up to this point the gamma function has been interpreted as purely real. Euler's great next leap was to extend it to complex values by changing variables. Thus, defining

$$u = \frac{x}{p+iq}$$

where p and q are real, positive constants results in

$$\Gamma(n) = \int_0^\infty e^{-(p+iq)u}\{(p+iq)u\}^{n-1}(p+iq)du$$

$$= \int_0^\infty (p+iq)^n \, u^{n-1} \, e^{-pu} \, e^{-iqu} \, du.$$

There is just a *little* cheating here, I must tell you. That is, while the original $\Gamma(n)$ integration path is along the real x-axis, the integration path of the transformed integral is along the u-line in the complex plane at angle $\alpha = -\tan^{-1}(q/p)$ to the real axis. I'll ignore this "subtle" concern here because such formal symbolic manipulations do lead to correct results, but later, when we get to Cauchy's theory of contour integration in the complex plane, I'll be *much* more careful about just where our integration paths are located. It was just this sort of devil-may-care looseness with complex changes of variable in integrals by Euler that motivated Cauchy's work, in fact.

Now, after changing the dummy variable of integration from u back to x, for no particular reason other than consistency of notation, we arrive at

$$\int_0^\infty x^{n-1} e^{-px} e^{-iqx}\, dx = \frac{\Gamma(n)}{(p+iq)^n}.$$

Euler then wrote $p + iq$ in polar form, i.e., as

$$p + iq = r \angle \alpha = r\{\cos(\alpha) + i\sin(\alpha)\} = re^{i\alpha},$$

where

$$r = \sqrt{p^2 + q^2}, \quad \alpha = \tan^{-1}\left(\frac{q}{p}\right).$$

Thus,

$$\int_0^\infty x^{n-1} e^{-px} e^{-iqx}\, dx = \frac{\Gamma(n)}{r^n e^{in\alpha}} = \frac{\Gamma(n)}{r^n} e^{-in\alpha}.$$

Finally, using Euler's identity to expand e^{-iqx} and $e^{-in\alpha}$ into real and imaginary parts, and equating the real and imaginary parts of both sides of the last equation, we have

$$\int_0^\infty x^{n-1} e^{-px} \cos(qx)\, dx = \frac{\Gamma(n)}{r^n} \cos(n\alpha),$$

$$\int_0^\infty x^{n-1} e^{-px} \sin(qx)\, dx = \frac{\Gamma(n)}{r^n} \sin(n\alpha).$$

Euler was now within sight of his original goal of evaluating the integrals I_1 and I_2. To this end he set $n = \frac{1}{2}$, $p = 0$ and $q = 1$. Since this gives $\alpha = \tan^{-1}(1/0) = \tan^{-1}(\infty) = \pi/2$ and $r = 1$, then

$$\int_0^\infty \frac{\cos(x)}{\sqrt{x}}\,dx = \Gamma\left(\frac{1}{2}\right)\cos\left(\frac{\pi}{4}\right) = \frac{1}{\sqrt{2}}\Gamma\left(\frac{1}{2}\right),$$

$$\int_0^\infty \frac{\sin(x)}{\sqrt{x}}\,dx = \Gamma\left(\frac{1}{2}\right)\sin\left(\frac{\pi}{4}\right) = \frac{1}{\sqrt{2}}\Gamma\left(\frac{1}{2}\right).$$

Now, as I showed earlier in this section, $\Gamma(\frac{1}{2}) = \sqrt{\pi}$ and so

$$\int_0^\infty \frac{\cos(x)}{\sqrt{x}}\,dx = \sqrt{\frac{\pi}{2}} = \int_0^\infty \frac{\sin(x)}{\sqrt{x}}\,dx.$$

But if you make the change of variable $x = s^2$ in the original I_1 and I_2 integrals, you should be able to quickly show that

$$I_1 = \frac{1}{2}\int_0^\infty \frac{\sin(x)}{\sqrt{x}}\,dx,$$

$$I_2 = \frac{1}{2}\int_0^\infty \frac{\cos(x)}{\sqrt{x}}\,dx,$$

and so $I_1 = I_2 = \frac{1}{2}\sqrt{\pi/2}$. What a genius Euler was—but he wasn't through yet!

Sometime between 1776 and his death in 1783, Euler showed, using complex numbers, that

$$\int_0^\infty \frac{\sin(x)}{x}\,dx = \frac{\pi}{2}.$$

This is an important integral, appearing over and over again all through physics and mathematics and in the theory of the electronic transmission of information. This wonderful calculation was published only years after Euler's death. What I'll actually show you here is a way to derive a somewhat more general result, also using complex numbers, with Euler's result as a special case.

Consider the complex integral $\int_0^\infty e^{(-p+iq)x}dx$, with $p > 0$ to insure the integral exists. This is easily integrated to give the formal result $(p + iq)/(p^2 + q^2)$. If Euler's identity is used to separate both sides of the result into their real and imaginary parts, then

$$\int_0^\infty e^{-px}\cos(qx)\,dx = \frac{p}{p^2+q^2},$$

$$\int_0^\infty e^{-px}\sin(qx)\,dx = \frac{p}{p^2+q^2}.$$

Fixing our attention on the first of these two statements, let us next integrate both sides with respect to q, i.e., let us treat q as a variable and p as a constant. Thus, with a and b arbitrary limits,

$$\int_a^b \left\{ \int_0^\infty e^{-px}\cos(qx)\,dx \right\} dq = \int_a^b \frac{p}{p^2+q^2}\,dq = \frac{1}{p}\int_a^b \frac{dq}{1+\left(\dfrac{q}{p}\right)^2}.$$

Next, reversing the order of integration in the double integral on the left, we have

$$\int_0^\infty e^{-px}\left\{ \int_a^b \cos(qx)\,dq \right\} dx = \int_0^\infty e^{-px}\left\{ \frac{\sin(qx)}{x}\Big|_a^b \right\} dx$$

$$= \int_0^\infty e^{-px}\,\frac{\sin(bx)-\sin(ax)}{x}\,dx.$$

To justify the reversal of the order of integration, it is necessary to demonstrate that the integrand obeys certain conditions on continuity and convergence, a result shown in any good book on calculus. For this book, which has no pretensions to being a rigorous textbook, I'll skip such a demonstration. But be assured, the reversal *is* okay here. The single integral on the right-hand side of the previous equation is also easy to integrate; simply make the change of variable $u = q/p$ [and so $du = (1/p)dq$] and write

$$\frac{1}{p}\int_a^b \frac{dq}{1+\left(\dfrac{q}{p}\right)^2} = \int_{a/p}^{b/p} \frac{du}{1+u^2} = \tan^{-1}(u)\Big|_{a/p}^{b/p} = \tan^{-1}\left(\frac{b}{p}\right) - \tan^{-1}\left(\frac{a}{p}\right).$$

If we set $b = 0$ in these two (equal) expressions we have

$$\int_0^\infty e^{-px}\,\frac{\sin(ax)}{x}\,dx = \tan^{-1}\left(\frac{a}{p}\right).$$

Letting $p \to 0$, we have

$$\lim_{p\to 0}\int_0^\infty e^{-px}\,\frac{\sin(ax)}{x}\,dx = \tan^{-1}(\pm\infty) = \begin{cases} \dfrac{\pi}{2} & \text{if } a > 0, \\[2mm] -\dfrac{\pi}{2} & \text{if } a < 0. \end{cases}$$

Finally, assuming we can interchange the limiting and integration operations, we arrive at

$$\int_0^\infty \frac{\sin(ax)}{x}\, dx = \begin{cases} \dfrac{\pi}{2} & \text{if } a > 0, \\ 0 & \text{if } a = 0, \\ -\dfrac{\pi}{2} & \text{if } a < 0, \end{cases}$$

a result called Dirichlet's discontinuous integral (the discontinuity is at $a = 0$), after the German mathematician Gustav Peter Lejeune Dirichlet (1805–59), who was Gauss' successor to the mathematics professorship at Göttingen upon Gauss' death in 1855. When he had finished his calculation of this integral for the special case of $a = 1$ Euler was justifiably proud, noting that "up to the present [it] has defeated all known artifices of calculation."

6.13 EULER'S REFLECTION FORMULA FOR $\Gamma(n)$, AND THE FUNCTIONAL EQUATION FOR $\zeta(n)$

In this section I will show you how the complex numbers can be used to derive a result due to Euler, and another due to Riemann, that are among the most famous identities in mathematics. Recall from the previous section that the gamma integral is defined as

$$\Gamma(n) = \int_0^\infty e^{-x} x^{n-1}\, dx.$$

As I did in that section, make the change of variable $x = t^2$ and so

$$\Gamma(n) = \int_0^\infty e^{-t^2} t^{2(n-1)}\, 2t\, dt,$$

from which it immediately follows by replacing n with $1 - n$ that

$$\Gamma(1-n) = \int_0^\infty e^{-t^2} t^{-2n}\, 2t\, dt.$$

The product $\Gamma(n)\Gamma(1 - n)$ is particularly interesting because of the property $\Gamma(n) = (n - 1)!$ That is,

$$n\Gamma(n)\Gamma(1 - n) = (-n)!\, (n!),$$

and if we could evaluate the product $\Gamma(n)\Gamma(1 - n)$ then we would have an expression from which we could calculate $(-n)!$ from $n!$, where $n \geq 0$.

So, let's write

$$\Gamma(n)\Gamma(1-n) = \left\{ \int_0^\infty e^{-x^2} x^{2(n-1)}\, 2x\, dx \right\}\left\{ \int_0^\infty e^{-y^2} y^{-2n}\, 2y\, dy \right\}$$

$$= 4\int_0^\infty \int_0^\infty e^{-(x^2+y^2)} x^{2n-1}\, y^{-(2n-1)}\, dx\, dy,$$

where I have used different symbols, x and y, for the dummy variables of integration to avoid subsequent confusion. Now, there is no doubt that this is an expression that makes most people weak in the knees on first glance. But its apparent difficulty melts away when attacked with the same ingenious trick I used in the previous section to calculate $\Gamma(\frac{1}{2})$.

So, as before, let us change to polar coordinates and write $x = r\cos(\theta)$ and $y = r\sin(\theta)$, with the corresponding differential area now $r\,dr\,d\theta$. To cover the first quadrant we require, as before, $0 \le r < \infty$ and $0 \le \theta \le \pi/2$. Since

$$x^2 + y^2 = r^2,$$

$$x^{2n-1}\, y^{-(2n-1)} = \left(\frac{x}{y}\right)^{2n-1} = \{\cot(\theta)\}^{2n-1},$$

then we have

$$\Gamma(n)\Gamma(1-n) = 4\int_0^\infty \int_0^{\frac{\pi}{2}} e^{-r^2} \{\cot(\theta)\}^{2n-1}\, r\, dr\, d\theta$$

$$= 4\int_0^\infty r\, e^{-r^2}\, dr \int_0^{\frac{\pi}{2}} \{\cot(\theta)\}^{2n-1}\, d\theta.$$

The r-integral is again elementary, i.e.,

$$\int_0^\infty r e^{-r^2}\, dr = -\frac{1}{2} e^{-r^2}\bigg|_0^\infty = \frac{1}{2},$$

and so

$$\Gamma(n)\Gamma(1-n) = 2\int_0^{\frac{\pi}{2}} \{\cot(\theta)\}^{2n-1}\, d\theta.$$

Now make a new change of variable, $s = \cot(\theta)$, which transforms the integral into

$$\Gamma(n)\Gamma(1-n) = 2\int_0^\infty \frac{s^{2n-1}}{1+s^2}\,ds,$$

or, if we write $2n = \alpha$,

$$\Gamma(n)\Gamma(1-n) = 2\int_0^\infty \frac{s^{\alpha-1}}{1+s^2}\,ds,$$

This is not a trivial integral but, in chapter 7, I'll show you how to do integrals like this one using De Moivre's theorem from chapter 3, combined with the nineteenth-century invention of the theory of integration in the complex plane. In particular, you'll see that

$$\int_0^\infty \frac{s^{\alpha-1}}{1+s^\beta}\,ds = \frac{\pi}{\beta \sin\left(\dfrac{\alpha}{\beta}\pi\right)},$$

and so for our problem here, with $\beta = 2$, we have the surprising, unexpected, indeed *astonishing* result that

$$\Gamma(n)\Gamma(1-n) = \frac{\pi}{\sin(n\pi)}.$$

This beautiful identity, due to Euler (1771), is called the *reflection formula* for the gamma function.

For $n = \frac{1}{2}$ we have $\Gamma^2(\frac{1}{2}) = \pi$, or $\Gamma(\frac{1}{2}) = \sqrt{\pi}$, which was calculated directly in the previous section during the analysis of the Fresnel integrals. From the first part of this section we also have the elegant-appearing expression

$$(-n)!(n!) = \frac{n\pi}{\sin(n\pi)}.$$

For example,

$$\left(-2\frac{1}{2}\right)! = \frac{\left(2\frac{1}{2}\right)\pi}{\left(2\frac{1}{2}\right)!\sin\left(2\frac{1}{2}\pi\right)} = \frac{\dfrac{5}{2}\pi}{\left(2\frac{1}{2}\right)\left(1\frac{1}{2}\right)\left(\dfrac{1}{2}\right)!}$$

$$= \frac{\dfrac{5}{2}\pi}{\dfrac{5}{2}\times\dfrac{3}{2}\times\dfrac{1}{2}\sqrt{\pi}} = \frac{4}{3}\sqrt{\pi} = 2.36327\dots.$$

In 1859 Riemann showed that there is an equally elegant formulation for the product $\Gamma(s)\zeta(s)$, i.e., for the product of the gamma and the zeta functions.

(It is traditional to write s for the complex variable in this analysis, rather than z. This, on the surface, appears to be a trivial point—it's just notation, after all—but in fact there *is* a deeper issue involved, and I will elaborate on it when we get to chapter 7.) Now, from the definition of the gamma function it follows that, if you make the change of variable $u = nx$,

$$\int_0^\infty e^{-nx} x^{s-1} \, dx = \frac{\Gamma(s)}{n^s}.$$

Then, summing over both sides,

$$\sum_{n=1}^\infty \int_0^\infty e^{-nx} x^{s-1} \, dx = \sum_{n=1}^\infty \frac{\Gamma(s)}{n^s} = \Gamma(s) \sum_{n=1}^\infty \frac{1}{n^s} = \Gamma(s)\zeta(s).$$

If we assume that we can interchange the integration and summation operations on the far left (be bold!), then we have

$$\Gamma(s)\zeta(s) = \int_0^\infty x^{s-1} \sum_{n=1}^\infty e^{-nx} \, dx$$

and, since the sum is just a geometric series, it is easy to do. The result is

$$\Gamma(s)\zeta(s) = \int_0^\infty \frac{x^{s-1}}{e^x - 1} \, dx.$$

So, just as with the product $\Gamma(n)\Gamma(1-n)$, for this new product we also arrive at a rather "non–freshman calculus looking" integral. And, just as before, the integral can be done by complex plane integration, and that is exactly how Riemann actually did it. I'll come back to this integral in section 7.8 but, not to leave you unsatisfied here, the result that Riemann arrived at is the wonderfully mysterious-appearing functional equation for the zeta function, one of the crown jewels of mathematics:

$$\zeta(s) = \zeta(1-s)\Gamma(1-s)2^s \pi^{s-1} \sin(\tfrac{1}{2}\pi s).$$

Here is a pretty application of the functional equation. Let $s = -2n$, where n is a non-negative integer. Then,

$$\zeta(-2n) = -\zeta(1+2n)\Gamma(1+2n)2^{-2n}\pi^{-(2n+1)}\sin(n\pi) = 0$$

because all the factors on the right are finite positive numbers except for $\sin(n\pi)$ which is, of course, zero. Thus, all the even, negative integers are zeros of the zeta function, as mentioned in section 6.4. We must exclude $n = 0$, however, as for that case $\zeta(1+2n) = \zeta(1) = S_1 = \infty$, and it can be shown that this infinity is sufficient to cancel the zero of sin(0). Indeed, it can be shown that $\zeta(0) = -\frac{1}{2}$.

Remember—the gamma reflection formula and the zeta functional equation both follow *if* you can do those two non–freshman calculus looking integrals, and you will be able to do them once you see how complex integrals are done in chapter 7.

The Nineteenth Century, Cauchy, and the Beginning of Complex Function Theory

7.1 INTRODUCTION

With the completion of the previous chapter we have really, I think, done pretty much everything we can do with just the imaginary $\sqrt{-1}$ itself, and its extension to complex numbers. To continue on, the next logical step is to consider *functions* of variables that are complex valued, i.e., functions $f(z)$ where $z = x + iy$. But then the question is, where do I stop? There is, today, an absolutely enormous literature in complex function theory, much of it purely mathematical, and just as much again of a practical, applications-oriented nature. Physicists and engineers, I believe, probably use complex variables and function theory more than do mathematicians. This book is not a textbook, but to jump into all of what we could jump into would turn this book into a *huge* textbook, perhaps one with a couple of thousand pages. You wouldn't buy it, and I certainly don't believe I would live long enough to write it.

So what I am going to do is show you just the very beginnings of modern complex function theory, complex integration, as initiated by the French genius Augustin-Louis Cauchy (1789–1857) in an 1814 memoir to the French Académie des Sciences.[1] I am going to do this for three reasons. First, as I've already stated, we can't do everything in a book like this one. Second, to begin at the beginning is really what I should do in a book that has a strong historical flavor. And third, I'm going to tell you about Cauchy's initial work simply because it is so beautiful.

When I first went off to Stanford to study electrical engineering, a story I told in the preface, I naturally began to take all the usual engineering math courses, such as differential and integral calculus, ordinary and partial differential equations, and even set theory. All of these courses were interesting to me, and in each one I quickly came to appreciate the power I was gaining in my ability to solve difficult problems. But it wasn't until my first course in complex function theory, in the autumn of 1960, that I experienced a totally new emotion—the pure pleasure of learning mathematics that was, in and of itself, "pretty."

Everything in that course snapped together crisply, cleanly, and with no spaces between the pieces, just like a well-made if complex puzzle. Things fit together in calculus, too, certainly, but there is a certain utilitarian feeling to all of it—at least there is for me, which perhaps is a character flaw. In complex function theory, however, the fundamental theorems are not only powerful in their generality but they are also *surprising*. For me, complex function theory was a revelation bordering on a mystical experience—another character flaw?

The part of complex function theory that fascinated me the most was its very first historical application, integration in the complex plane, or what is called *contour integration*. With Cauchy's theory of complex integration one could, almost without effort, calculate the values of a seemingly endless number of incredibly odd, strange, and downright wonderfully mysterious-looking definite real integrals. I learned how to show, for example, that

$$\int_{-\infty}^{\infty} \frac{\cos(x)}{1+x^2}\, dx = \frac{\pi}{e}.$$

Such calculations were to me, then, seemingly possible only if one had the powers of a sorcerer. But I was wrong—the theory behind such calculations was actually developed by a man no more sinister than a quirky, devout Christian supporter of the traditional line of French kings.

7.2 Augustin-Louis Cauchy

When the Reign of Terror swept Paris in 1793 Cauchy's father, a high-ranking government official, fled with his family to a countryside village where he found other intellectuals who had also gotten out of the city until the heads stopped falling and the blood stopped flowing. One of the neighbors who also narrowly escaped the guillotine, for example, was Laplace, who was as great a political survivor as he was a mathematical physicist. The temporary, hasty move may even have been stimulating, as Cauchy evidently gave early indications of his genius. The great French mathematician, Lagrange, for example, is said to have told Cauchy's father that his son would one day be a scientific star, but warned against letting the boy see an advanced mathematical book before reaching seventeen!

His early training was as a civil engineer, and by 1810 he was involved with the construction of a naval base from which Napoleon intended to launch attacks against England. For unknown reasons of health—which may in fact have been due to the years of childhood malnutrition he suffered during his family's absence from Paris—Cauchy abandoned the physical rigors of the military engineering life in 1811 and began his spectacularly productive career as a mathematician. By the time he died he had written more than eight hundred papers and seven books, an output second only to Euler's.

The memoir of 1814 came just three years after he started his new career, and it would have been the pride of any mathematician twice his age. It was not seen that way at first, however, by the elders who reviewed it. They generally endorsed it—although Legendre, in particular, critized Cauchy's evaluation of an integral Legendre had just published in his new *Exercises du Calcul Intégral* because their answers were different (*Cauchy* was correct)—but they also failed to recognize that Cauchy was about to launch an entirely new branch of mathematics, that of complex function theory. This failure was almost certainly due to Cauchy's stated reasons for writing his memoir, which were twofold and somewhat mundane. First, he was attempting to explain why the value of an iterated integral might be different, depending on the order in which the individual integrations are performed, a difference he called the *residue*. And a second goal was to place on firmer ground the rather carefree way earlier mathematicians had evaluated certain definite integrals with tricks involving complex numbers, e.g., Euler's evaluation of the Fresnel integrals, discussed in chapter 6. The man who did all this was no one-dimensional personality, but rather was a highly complex individual.

Although born only weeks after the storming of the Bastille and the birth of the French Revolution, Cauchy was an almost fanatical supporter of the Bourbon dynasty all his life. When Lazare Carnot and Carnot's former mathematics professor, Gaspard Monge, were expelled from the Académie des Sciences in 1816 for political reasons, for example, Cauchy happily accepted a non-elected appointment in Monge's place. Monge had been an old crony of the usurper Napoleon, and Carnot's role, in particular, in voting for the execution of Louis XVI twenty-three years before, no doubt made this decision easy for Cauchy—it bothered him not at all to see a supporter of a false king, and a regicide, in disgrace.

As a person Cauchy seems to have been somewhat lacking in common sense, with a reputation for naive, childish, even boorish behavior. For example, when the July Revolution of 1830 put the Orleans "Citizen-King" of the bourgeoisie, Louis-Philippe, on the throne, an oath of allegiance was required for Cauchy to retain his three academic chairs at as many different institutions. Cauchy refused to so swear (he wasn't against kings and their class in general, just those who weren't Bourbon kings) and went into self-imposed exile along with the deposed Charles X. Perhaps he feared another bloody revolution, but it does seem odd that he left his family behind for four years. He finally returned to France in 1838. And earlier, when the Norwegian mathematician Niels Abel visited Cauchy in 1826, he wrote that he found the man a religious "bigot" and worse. This is not to say Cauchy was a "terrible" person. It seems improbable that a man who could write the following love poem to his eventual wife could really be all that bad:

I shall love you, my tender friend,
Until the end of my days;
And since there is another life
Your Louis will love you always.

But what does Cauchy's personality matter today?—it is his *mathematics* that is beautiful. Just hours before his death Cauchy had been talking with the Archbishop of Paris about his (Cauchy's) plans for charity work—yet another curious twist to his unusual character. The Archbishop recalled that Cauchy's last words to him were "Men pass away but their deeds abide." Cauchy's deeds will last as long as people study mathematics.

There is direct evidence from Gauss' own private notes and letters (predating 1814) that he knew much, if not all and more, of what is in Cauchy's 1814 memoir. But as was Gauss' style, he sat on it all until he could get everything "just right" and so the glory is rightfully all Cauchy's. As the nineteenth-century German mathematician Leopold Kronecker wrote long after both Cauchy and Gauss were dead, there is "a big difference between someone who publishes a mathematical proof and an indication of its complete scope, and another who only incidentally communicates it privately to a friend. Therefore the theorem [what is called the *first integral theorem* in section 7.5] can rightly be designated as the *Cauchy theorem*." But, of course, Cauchy did not do everything, and there was an already established foundation that he could build upon. For example, Cauchy's paper opens with the assumption that his readers understood, in essence, what is meant today by saying the complex function $f(z)$ is analytic in a region of the complex plane.[2] So let me start the last technical development of this book by explaining what that is all about.

7.3 ANALYTIC FUNCTIONS AND THE CAUCHY-RIEMANN EQUATIONS

To begin, I want to emphasize that x and y will be our most primitive variables, with both of them real. x defines the real axis, and y roams along the imaginary axis which is not just y, of course, but iy. The next step up in sophistication is the complex variable $z = x + iy$. And finally, when I write of a complex function I mean $f(z)$, e.g., $f(z) = f(x + iy) = z^2$, or $f(z) = f(x + iy) = e^z$. The function $f(z)$ has, of course, real and imaginary parts, which in turn are functions of x and y. For example,

$$f(z) = z^2 = (x + iy)^2 = x^2 - y^2 + i\,2xy$$

and

$$f(z) = e^z = e^{x+iy} = e^{\,x}\cos(y) + i\,e^{\,x}\sin(y).$$

In general, I'll write $f(z) = u(x,y) + iv(x,y)$ and so, in the case of $f(z) = z^2$, we have $u(x,y) = x^2 - y^2$ and $v(x,y) = 2xy$. Be very clear on this in your mind: both u and v are *real* functions of the *real* variables x and y.

The start of complex function theory is found in the answer to the following question: what is the *derivative* of $f(z)$? The formal answer is precisely what you would expect, recalling how the derivative of a function of a single real variable is defined in freshman calculus. So, formally, the answer is that if $f(z)$ has a derivative $f'(z)$ at $z = z_0$, then

$$f'(z_0) = \frac{df}{dz}\bigg|_{z=z_0} = \lim_{\Delta z \to 0} \frac{f(z_0 + \Delta z) - f(z_0)}{\Delta z}.$$

The vanishing of Δz is not so straightforward as it is in the case of a single real variable, however. In that simpler case, where we let $\Delta x \to 0$ to calculate $f'(x)$, Δx only has to vanish on the one-dimensional real axis because the single real variable x exists only on the real axis. But since z exists anywhere in the two-dimensional complex plane, then Δz has infinitely many more ways open to it to vanish. So how *does* $\Delta z \to 0$?

The answer is that we want to have the most condition-free definition possible for $f'(z)$, and so we will insist that it shouldn't matter how $\Delta z \to 0$. This is a philosophical position, of course. *Why* do we want the most condition-free definition? Well, in mathematics we are of course free to define anything any way we please, but it is always best when the particular definitions we do use turn out to be useful, i.e., allow us to solve difficult problems. It will turn out that defining $f'(z)$ this way will do exactly that for us. Still, there is no such thing as a free lunch, and by insisting on this freedom we will find that we have imposed some restrictions elsewhere. To anticipate matters, let me tell you that we will find that there are conditions on $u(x,y)$ and $v(x,y)$, the real and imaginary parts of $f(z)$, if $f'(z)$ is to exist at $z = z_0$ no matter how $\Delta z \to 0$. These conditions are given by the Cauchy-Riemann partial differential equations; the equations are

$$\frac{\partial u}{\partial x} = \frac{\partial v}{\partial y},$$

$$\frac{\partial u}{\partial y} = -\frac{\partial v}{\partial x},$$

evaluated at $z = z_o$.

The Cauchy-Riemann or C-R equations are necessary and almost sufficient conditions on u and v, where $f = u + iv$, for f to have a unique derivative at $z = z_o$. Here's how to show necessity. Write $z_o = x_o + iy_o$ and $\Delta z = \Delta x + i\Delta y$. Then, $\Delta z \to 0$ is equivalent to requiring both $\Delta x \to 0$ and $\Delta y \to 0$. Thus,

$$f'(z_0) = \lim_{\substack{\Delta x \to 0 \\ \Delta y \to 0}} \frac{f(x_0 + \Delta x, y_0 + \Delta y) - f(x_0, y_0)}{\Delta x + i\,\Delta y}.$$

Now, out of all of the infinity of ways that both Δx and Δy can vanish, let us consider just two. First, let us assume $\Delta y = 0$ and so $\Delta z = \Delta x$, i.e., z approaches z_o parallel to the real axis. Second, let us assume $\Delta x = 0$ and so $\Delta z = i\Delta y$, i.e., z approaches z_o parallel to the imaginary axis. If $f'(z_o)$ is to be unique, independent of the details of just how $\Delta z \to 0$, then these two particular results must certainly be equal.

In the first case we have

$$
\begin{aligned}
f'(z_0) &= \lim_{\Delta x \to 0} \frac{f(x_0 + \Delta x, y_0) - f(x_0, y_0)}{\Delta x} \\
&= \lim_{\Delta x \to 0} \frac{\{u(x_0 + \Delta x, y_0) + iv(x_0 + \Delta x, y_0)\} - \{u(x_0, y_0) + iv(x_0, y_0)\}}{\Delta x} \\
&= \lim_{\Delta x \to 0} \frac{\{u(x_0 + \Delta x, y_0) - u(x_0, y_0)\} + i\{v(x_0 + \Delta x, y_0) - v(x_0, y_0)\}}{\Delta x} \\
&= \frac{\partial u}{\partial x} + i\frac{\partial v}{\partial x}.
\end{aligned}
$$

In the second case we have

$$
\begin{aligned}
f'(z_0) &= \lim_{i\Delta y \to 0} \frac{f(x_0, y_0 + \Delta y) - f(x_0, y_0)}{i\Delta y} \\
&= \lim_{i\Delta y \to 0} \frac{\{u(x_0, y_0 + \Delta y) + iv(x_0, y_0 + \Delta y)\} - \{u(x_0, y_0) + iv(x_0, y_0)\}}{i\Delta y} \\
&= \lim_{i\Delta y \to 0} \frac{\{u(x_0, y_0 + \Delta y) - u(x_0, y_0)\} + i\{v(x_0, y_0 + \Delta y) - v(x_0, y_0)\}}{i\Delta y} \\
&= \frac{1}{i}\frac{\partial u}{\partial y} + \frac{\partial v}{\partial y} = \frac{\partial v}{\partial y} - i\frac{\partial u}{\partial y}.
\end{aligned}
$$

Equating the real and imaginary parts of these two expressions for $f'(z_0)$ gives the two C-R equations.

That analysis shows necessity, but to show sufficiency we would next have to show that *any* other way that Δz might vanish, not just the two ways considered here, leads to the same $f'(z_0)$. This isn't a terribly hard thing to do, but for that I'll refer you to any good textbook on complex function theory. Here, we'll take it on faith that if the C-R equations are satisfied and if the partial derivatives are continuous then the function $f(z)$ is analytic.

Riemann, of zeta function fame, was the first to use the mathematical argument I've presented here, in his 1851 doctoral dissertation. It is now the standard textbook approach. In fact, however, the C-R equations had been discovered from physical considerations long before either Cauchy or Riemann. In 1752 d'Alembert arrived at expressions equivalent to the C-R equations through a study in hydrodynamics, specifically the determination of necessary conditions for a spinning mass of fluid to be in equilibrium. Arguing that such a fluid both is incompressible and has no internal currents led d'Alembert to the equations. But that, of course, is physics and not the pure mathematics that a mathematician wants.

If $f(z) = u + iv$ satisfies the C-R equations, not only at $z = z_0$ but at all points in a *region* (or *domain* or *neighborhood*) surrounding $z = z_0$, then $f(z)$ is said to be *analytic* in that region. To pick perhaps the simplest nontrivial $f(z)$ as an example, suppose $f(z) = z = x + iy$. Thus, $u = x$ and $v = y$, and so

$$\frac{\partial u}{\partial x} = 1, \quad \frac{\partial v}{\partial y} = 1,$$

$$\frac{\partial u}{\partial y} = 0, \quad \frac{\partial v}{\partial x} = 0,$$

and obviously the C-R equations are satisfied for all x and y, i.e., $f(z) = z$ is analytic in the domain of the entire finite complex plane; indeed, such a function is called an *entire* function. The caveat of "finite" is important, as $|f(z)|$ blows up as $|z| \to \infty$. So $f(z) = z$ certainly cannot be analytic at infinity. In fact, there is a theorem that states the only complex functions that are entire over the *infinite* plane are *constants*—all four partial derivatives then are identically zero.

Not all $f(z)$ are analytic. To see this, suppose now, for a second example, that $f(z) = \bar{z} = x - iy$. Then, $u = x$ and $v = -y$, and so

$$\frac{\partial u}{\partial x} = 1, \quad \frac{\partial v}{\partial y} = -1,$$

and obviously one of the C-R equations is never satisfied, i.e., $f(z) = \bar{z}$ is analytic in no region of the complex plane. It is astonishing, I think, that such an apparently minor change in $f(z)$ makes such an enormous difference in the analyticity of the function.

Let me now show you an implication of the C-R equations that hints at the value of complex function theory in physics and engineering. If you differentiate the C-R equations to form all possible second derivatives, then you have

$$\frac{\partial^2 u}{\partial y \partial x} = \frac{\partial^2 v}{\partial y^2},$$

$$\frac{\partial^2 v}{\partial y \partial x} = -\frac{\partial^2 u}{\partial y^2},$$

$$\frac{\partial^2 u}{\partial x^2} = \frac{\partial^2 v}{\partial x \partial y},$$

$$\frac{\partial^2 v}{\partial x^2} = -\frac{\partial^2 u}{\partial x \partial y}.$$

The first and last expressions give

$$\frac{\partial^2 v}{\partial x^2} + \frac{\partial^2 v}{\partial y^2} = 0$$

and the middle two give

$$\frac{\partial^2 u}{\partial x^2} + \frac{\partial^2 u}{\partial y^2} = 0.$$

Both of these equations are *Laplace's equation*—after the French mathematical physicist Pierre Simon de Laplace (1749–1827), who was Cauchy's neighbor during the Terror—which occurs in such diverse applications as fluid dynamics, electrostatics, and optics. The fact that the real and the imaginary parts of every analytic function each are solutions to Laplace's equation can be used to solve many important problems with physical origin. This was first shown in Riemann's seminal doctoral dissertation. The real and imaginary parts, u and v, are called *harmonic conjugates* of each other.

Analytic functions are clearly a rather special subset of all possible complex functions, but certain broad classes of functions do make the cut. They include:

1. Every polynomial of z is analytic.
2. Every sum and product of two analytic functions is analytic.
3. Every quotient of analytic functions is analytic except at those values of z where the denominator is zero.
4. An analytic function of an analytic function is analytic.

Thus, from (1) $f(z) = z^2$ and $f(z) = e^z$ are analytic: the first case because it is a polynomial and the second case because the exponential can be expanded in a power series polynomial. From (2) $f(z) = z^2 e^z$ is analytic, and from (3) $f(z) = e^z/(z^2 + 1)$ is analytic except at $z = \pm i$, which are called the *singularities* of $f(z)$ because, for those values of z, $f(z)$ is infinite. Historically, an analytic function was taken to be any function that can be expanded as a power series.

7.4 CAUCHY'S FIRST RESULT

Suppose we start with the function $f(z) = e^{-z^2}$. If you work out the real and imaginary parts of this function you can show with little difficulty first that

$$u(x,y) = e^{-x^2} e^{y^2} \cos(2xy),$$
$$v(x,y) = -e^{-x^2} e^{y^2} \sin(2xy),$$

and then that these two functions satisfy the C-R equations for all finite x and y. That is, $f(z)$ is analytic everywhere in the finite complex plane and so is an entire function. Now, what Cauchy did in his 1814 memoir was to calculate the integral of this $f(z)$ around the closed rectangular path shown in figure 7.1,

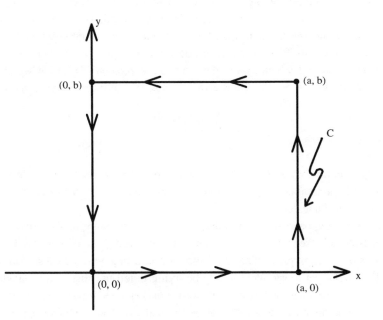

Figure 7.1. The rectangular contour from Cauchy's 1814 memoir.

an expression I'll write as $\oint_C f(z)dz$. The C denotes the rectangular *contour* that is the path of integration. The circle through the integral sign means the contour is *closed*. Such integrals are often called line or contour integrals. "Contour" is usually reserved for paths that are closed, while "line" may or may not mean that the start and end of the integration path are the same point.

The notation of a contour integral may look a little odd, but as you'll soon see the actual mechanics of calculating such expressions is not at all difficult. You should not be surprised that, when writing the integrals of complex-valued functions, we have to specify the path C. This is because unlike real functions, which have their integration intervals confined to the real axis, complex functions have the entire complex plane available to the variable z over which to roam. This additional freedom does bring some new problems with it—and I'll show you some when we get to them—but it also opens up tremendous new possibilities, too. As I will show you, for example, simply picking different C's for the same $f(z)$ can lead to quite different conclusions.

A famous example of this possibility was discussed in 1815 by the French mathematician Siméon-Denis Poisson (1781–1840) and published in 1820, in the form of the integral $I = \int_{-1}^{1} dx/x$. By a symmetry argument one is tempted to say this integral is zero because $1/x$ is an odd function, i.e., there is infinite negative area from -1 to 0 and infinite positive area from 0 to 1. This leaves an uncomfortable feeling of pulling a swindle, however, because $\infty + (-\infty)$ can be anything, not just zero.[3] A problem is caused, as it turns out, by integrating along the real axis right through a point—the origin—where the integrand blows up. Poisson, however, observed that there is another way to get from -1 to $+1$ which avoids the integrand explosion. That is, if we change variables to $x = -e^{i\theta}$, with $0 \leq \theta \leq \pi$ radians, then the integral becomes

$$I = \int_{0}^{\pi} \frac{-ie^{i\theta}d\theta}{-e^{i\theta}} = \int_{0}^{\pi} i\, d\theta = i\pi.$$

The integrand is always well defined now because the integration path swings around the fatal point $x = 0$, along a semicircular arc in the upper half of the complex plane.

In what appears to be an unintentional touch of humor in Kolmogorov and Yushkevich (see note 1) on Poisson's idea, we read "It is obvious that Poisson was on the right path." The calculation is interesting, yes, indeed so, but it is a little early yet to see what it may be trying to tell us. Historically, however, this is particularly interesting because Poisson's 1820 paper represents the first time an integration path not on the real axis had explicitly appeared in print; Cauchy's 1814 memoir was not published until 1825. I should add, however, that Poisson had seen Cauchy's 1814 memoir when it was originally submitted to the French Academy.

Now, let's get down to the real business at hand. Why did Cauchy calculate his particular contour integral? Surprisingly, it wasn't to learn the answer—he already knew it was zero. He knew this because the contour integral around *any* non-self-intersecting closed path C, of a function that is analytic everywhere inside and on C, is always zero. That is,

$$\oint_C f(z)dz = 0$$

if C and its interior is an analytic domain of $f(z)$. This astonishing result is Cauchy's first integral theorem, and I will show you its beautiful proof in just a bit. For the rest of this section let's assume the truth of the theorem.

So, you probably are asking again, *why* did Cauchy bother to calculate this contour integral if he already knew the result had to be zero? The answer is that, yes, the whole integral is zero, but we will be able to break it up into several parts and, as you know, several different quantities that are not zero can add to zero. By doing the contour integral Cauchy could calculate the values of the several parts, and it is the values of the parts that we are going to be especially interested in knowing. The calculation using the particular $f(z)$ that opened this section was done right at the beginning of Cauchy's 1814 paper, and the result is a generalization of the famous integral that I discussed in section 6.12 that Lord Kelvin was so enthusiastic about. Here's what Cauchy did, in essence.

We start by writing

$$\oint_C f(z)dz = \oint_C (u+i\ v)(dx+i\ dy)$$
$$= \oint_C (u\ dx - v\ dy) + i\ \oint_C (v\ dx + u\ dy) = I_1 + i\ I_2,$$

with

$$I_1 = \oint_C (u\ dx - v\ dy) = 0,$$
$$I_2 = \oint_C (v\ dx + u\ dy) = 0.$$

I_1 and I_2 must each be zero because a complex quantity $I_1 + iI_2$ is zero only if both real and imaginary parts vanish separately. Now, recalling the expressions for u and v at the start of this section and looking at figure 7.1 again, we can write the following as we start at the origin and travel counterclockwise around C—there are obviously two ways to integrate around a closed curve, and by convention the standard choice is CCW:

$(0,0) \rightarrow (a,0)$: $y = 0$ and $dy = 0$, and so $u = e^{-x^2}$, $v = 0$.
$(a,0) \rightarrow (a,b)$: $x = a$ and $dx = 0$, and so $u = e^{-a^2} e^{y^2} \cos(2ay)$,
$\quad v = -e^{-a^2} e^{y^2} \sin(2ay)$.
$(a,b) \rightarrow (0,b)$: $y = b$ and $dy = 0$, and so $u = e^{-x^2} e^{b^2} \cos(2bx)$,
$\quad v = -e^{-x^2} e^{b^2} \sin(2bx)$.
$(0,b) \rightarrow (0,0)$: $x = 0$ and $dx = 0$, and so $u = e^{y^2}$, $v = 0$.

Writing out I_1 in detail, as we go around C, using the above expressions, we have I_1 as the sum of several real integrals:

$$I_1 = \int_0^a e^{-x^2} \, dx + \int_0^b e^{-a^2} e^{y^2} \sin(2ay) \, dy + \int_a^0 e^{-x^2} e^{b^2} \cos(2bx) \, dx = 0,$$

or, rearranging and remembering that $\int_a^0 = -\int_0^a$,

$$\int_0^a e^{-x^2} \, dx = e^{b^2} \int_0^a e^{-x^2} \cos(2bx) dx - e^{-a^2} \int_0^b e^{y^2} \sin(2ay) \, dy.$$

Now, let $a \rightarrow \infty$, i.e., let C become an infinitely wide rectangle. Since the height b remains finite, then the rightmost integral is finite but its coefficient of e^{-a^2} goes to zero, and so we end up with

$$\int_0^\infty e^{-x^2} \, dx = e^{b^2} \int_0^\infty e^{-x^2} \cos(2bx) \, dx$$

or

$$\int_0^\infty e^{-x^2} \cos(2bx) \, dx = e^{-b^2} \int_0^\infty e^{-x^2} \, dx.$$

But the integral on the right is just Lord Kelvin's favorite integral from section 6.12, equal to $\frac{1}{2}\sqrt{\pi}$, and so we have

$$\int_0^\infty e^{-x^2} \cos(2bx) \, dx = \frac{1}{2} e^{-b^2} \sqrt{\pi},$$

a *generalization* of Lord Kelvin's favorite integral (to which it reduces when $b = 0$).[4]

Isn't that *incredible?* The result just seems to come out of nowhere. Of course, it really doesn't—Cauchy's first integral theorem is what gave Cauchy the muscle to do this wonderful calculation. Now, what about I_2, what is *it* equal to? You can go through the details of I_2 just as I did for I_1 and show, just as Cauchy did in his 1814 memoir, that the I_2 integral leads to the result

$$\int_0^\infty e^{-x^2} \sin(2bx) \; dx = e^{-b^2} \int_0^b e^{y^2} \, dy.$$

This reduces to the trivial $0 = 0$ when $b = 0$, but other than that we simply have one integral in terms of another integral. It's a true statement, sure, but not nearly as useful a result as the I_1 integral. To evaluate the right-hand side of this identity requires a power series expansion, which was worked out by George Stokes in an 1857 paper.[5]

Now, to really dazzle you, let me show you how simply changing the contour C from the rectangle of figure 7.1 to the pie-shaped wedge of figure 7.2 will result in a pretty derivation of the Fresnel integrals that, following Euler's lead, I did in chapter 6. As before, $f(z) = e^{-z^2}$. The wedge contour is a chunk out of a circular pie with radius R. Since $0 \le \theta \le \pi/4$ radians, it is more

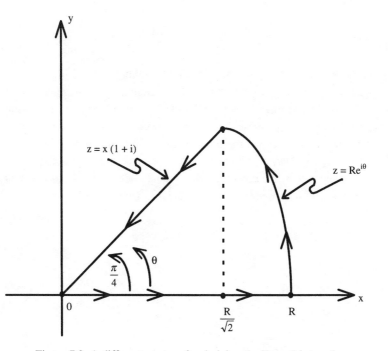

Figure 7.2. A different contour for deriving the Fresnel integrals.

precisely one-eighth of the pie. Now, along the real axis we have $z = x$ and so $dz = dx$. Along the circular arc I'll write z in polar form, as $z = Re^{i\theta}$, and so $dz = iRe^{i\theta}d\theta$. And on the final part of C that closes the integration path we have $z = (1 + i)x$ and so $dz = (1 + i)dx$. So, from Cauchy's first integral theorem, we can write

$$\oint_C f(z)dz = 0 = \int_0^R e^{-x^2}\ dx + \int_0^{\frac{\pi}{4}} e^{-R^2 e^{i2\theta}} i\,Re^{i\theta}\ d\theta + \int_{\frac{R}{\sqrt{2}}}^0 e^{-i2x^2}\ dx(1+i),$$

Now, let $R \to \infty$. As shown in box 7.1 the second integral vanishes as $R \to \infty$ and so

BOX 7.1

WHY $I = \lim_{R\to\infty} \int_0^{\pi/4} e^{-R^2 e^{i2\theta}}\ iRe^{i\theta}\ d\theta = 0$

$$|I| = \left|\int_0^{\pi/4} e^{-R^2 e^{i2\theta}} iRe^{i\theta}\ d\theta\right| \le \int_0^{\pi/4} \left|e^{-R^2 e^{i2\theta}}\right|\ \left|iRe^{i\theta}\right|\ d\theta$$

$$= \int_0^{\pi/4} \left|e^{-R^2\{\cos(2\theta)+i\sin(2\theta)\}}\right|\ R\ d\theta = R\int_0^{\pi/4} \left|e^{-R^2\cos(2\theta)}e^{-iR^2\sin(2\theta)}\right|\ d\theta.$$

Or, at last, $|I| \le R \int_0^{\pi/4} e^{-R^2\cos(2\theta)}d\theta$. Changing variables to $\phi = 2\theta$, we have $|I| \le \frac{1}{2}R \int_0^{\pi/2} e^{-R^2\cos(\phi)}d\phi$. From geometry we have $\cos(\phi) \ge (2/\pi)\pi/2 - \phi)$ for $0 \le \phi \le \pi/2$ radians. This is easily seen by simply graphing the quarter-cycle of $y = \cos(\phi)$ from $\phi = 0$ to $\pi/2$, on the same set of axes with a graph of the straight line $y = (2/\pi)\pi/2 - \phi) = 1 - (2/\pi)\phi$. The two graphs intersect at $\phi = 0$ and $\phi = \pi/2$ radians, with the cosine curve obviously above the straight line at all values of ϕ between 0 and $\pi/2$ radians. Thus,

$$|I| \le \frac{1}{2}R\int_0^{\pi/2} e^{-R^2\frac{2}{\pi}\left(\frac{\pi}{2}-\phi\right)}d\phi = \frac{1}{2}R\int_0^{\pi/2} e^{-R^2}e^{\frac{2R^2}{\pi}\phi}d\phi$$

$$= \frac{1}{2}Re^{-R^2}\int_0^{\pi/2} e^{\frac{2R^2}{\pi}\phi}d\phi = \frac{1}{2}Re^{-R^2}\left(\frac{\pi}{2R^2}e^{\frac{2R^2\phi}{\pi}}\right)\Big|_0^{\frac{\pi}{2}}$$

$$= \frac{\pi e^{-R^2}}{4R}\left(e^{R^2}-1\right) = \frac{\pi}{4R}\left(1-e^{-R^2}\right) \to 0 \text{ as } R \to \infty.$$

$$\int_0^\infty e^{-x^2} \; dx = \frac{1}{2}\sqrt{\pi} = \int_0^\infty e^{-i2x^2} \; (1+i) \; dx$$

$$= \int_0^\infty \{\cos(2x^2) - i\sin(2x^2)\} \; (1+i) \; dx.$$

Equating real and imaginary parts of the last integral to the purely real value of $\frac{1}{2}\sqrt{\pi}$ gives

$$\int_0^\infty \cos(2x^2) \; dx = \int_0^\infty \sin(2x^2) \; dx$$

and

$$\int_0^\infty \cos(2x^2) \; dx + \int_0^\infty \sin(2x^2) \; dx = \frac{1}{2}\sqrt{\pi}.$$

Thus,

$$\int_0^\infty \cos(2x^2) \; dx = \int_0^\infty \sin(2x^2) \; dx = \frac{1}{4}\sqrt{\pi}$$

or, upon changing variables to $s^2 = 2x^2$, this becomes

$$\int_0^\infty \cos(s^2) \; ds = \int_0^\infty \sin(s^2) \; ds = \frac{1}{4}\sqrt{2\pi} = \frac{1}{2}\sqrt{\frac{\pi}{2}},$$

just as Euler had deduced thirty years before.

All of this section depends on Cauchy's first integral theorem, which I have so far asked you simply to take on faith. Let me now demonstrate that your faith has not been given in vain.

7.5 CAUCHY'S FIRST INTEGRAL THEOREM

Given a complex function $f(z)$ that is analytic everywhere on the contour C and everywhere in the region R that C is the edge of, as shown in figure 7.3, Cauchy's first integral theorem says

$$\oint_C f(z)dz = 0.$$

In an 1811 letter to a friend Gauss wrote "This is a beautiful theorem whose simple proof I shall give on a suitable occasion." This he failed to do and so it

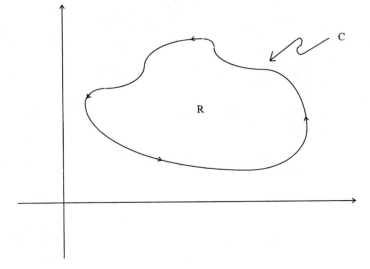

Figure 7.3. A simple curve, with an inside and an outside.

is Cauchy who gets the credit. By convention we take the direction of the contour integration, as I mentioned before, to be counterclockwise (CCW), i.e., as we move on C around R, R is always on our left. This defines the *interior* of C. If we travel on C in the opposite sense then the algebraic sign of the integral will be reversed.

The idea that a closed, non-self-intersecting curve, called a *simple* curve, separates an "inside" from an "outside" is obvious to most people, but it turns out to be not so easy to prove. Stated as a theorem it is called the Jordan curve theorem, after the French mathematician Camille Jordan (1838–1922), who offered an erroneous proof of it in 1887. The first correct proof did not appear until 1905. In this book all the simple curves we will consider are *so* simple, e.g., rectangles and circles, that I'll take the Jordan curve theorem to be obviously true. The interior of a simple curve is a *simply connected region*, which geometrically means every closed curve in the region, simple or not, encloses only points in the region. If a region is not simply connected then it is called multiply connected—a simple (oh, the limitations of prose!) example of a multiply connected region is a simply connected region that then has a hole cut in it. The points "in the hole" are then part of the *outside*.

To prove Cauchy's theorem, write

$$\oint_C f(z)dz = \oint_C \{u(x,y)+iv(x,y)\}\ \{dx+i\ dy\}$$

$$= \oint_C \{(u\ dx - v\ dy)+i(v\ dx+u\ dy)\}$$

$$= \oint_C (u\ dx - v\ dy)+i\oint_C (v\ dx+u\ dy).$$

The last two contour integrals separately vanish if the C-R equations hold, which they do because we have assumed that $f(z)$ is analytic. To prove this statement, I will invoke a famous result, called *Green's theorem,* which says that if $P(x,y)$ and $Q(x,y)$ are two real functions of x and y, then

$$\oint_C (Pdx+Qdy) = \iint_R \left(\frac{\partial Q}{\partial x} - \frac{\partial P}{\partial y} \right)\ dx\ dy.$$

Now, just as I did before for Cauchy's first integral theorem, I'm going to ask you to believe this result for just a bit. I will, I promise you, prove this important result soon—in the very next section.

In his 1814 memoir Cauchy did not use Green's theorem for the simple reason that Green hadn't published it yet. Cauchy used an entirely equivalent approach, but since Green's theorem is much more transparent that is what I'm using here. Cauchy clearly had the same opinion, because in a later presentation (1846) he adopted Green's theorem. When you look at the next section for the proof of Green's theorem you will see precisely what the details are of integrating over the region R (over the inside of C) as given by the above double integral on the right, but for our purposes here such details don't come into play. That's because for our first contour integral in Cauchy's theorem, $\oint_C(udx - vdy)$, we have $P = u$ and $Q = -v$ and so by Green's theorem

$$\oint_C (u\ dx - v\ dy) = \iint_R \left(-\frac{\partial v}{\partial x} - \frac{\partial u}{\partial y} \right)\ dx\ dy.$$

But the C-R equations tell us that $-\partial v/\partial x - \partial u/\partial y = 0$ and so

$$\oint_C (u\ dx - v\ dy) = 0.$$

Similarly, for the second contour integral in Cauchy's theorem, $\oint_C(vdx + udy)$, we have $P = v$ and $Q = u$ and so by Green's theorem

$$\oint_C (v\ dx + u\ dy) = \iint_R \left(\frac{\partial u}{\partial x} - \frac{\partial v}{\partial y} \right)\ dx\ dy.$$

But the C-R equations tell us that $\partial u/\partial x - \partial v/\partial y = 0$ and so

$$\oint_C (u\ dx + u\ dy) = 0,$$

which completes the proof of Cauchy's first integral theorem.

When you read the next section for the proof of Green's theorem you will see that I assume not only that $f'(z)$ exists, but also that it is continuous. It was not until 1884, many years after Cauchy's death, that the French mathematician Edouard Goursat (1858–1936) showed that the first integral theorem is true even if $f'(z)$ is not continuous. Two years later the Italian Giacinto Morera (1856–1909) showed that the converse to Cauchy's first integral theorem is also true. That is, if $f(z)$ is a continuous function in a simply connected region, and if for any simple curve C in that region $\oint_C f(z)dz = 0$, then $f(z)$ is analytic in the region.

7.6 GREEN'S THEOREM

The theorem used in the last section to prove Cauchy's first integral theorem,

$$\oint_C (Pdx + Qdx) = \iint_R \left(\frac{\partial Q}{\partial x} - \frac{\partial P}{\partial y} \right)\ dx\ dy,$$

where the contour C is the boundary edge of the two-dimensional region R in the complex plane, is named after the self-educated English mathematician George Green (1793–1841). This is the name the theorem generally goes under, although I have also seen various textbook authors call it Gauss' theorem or Stokes' theorem. These authors are all referring to the above statement, however, so why the multiplicity of names?

In fact, theorems relating contour, surface, and volume integrals were all the rage in the first half of the nineteenth century. As the historian of mathematics Jesper Lützen has written, "any mathematician working in [the theories of electricity or gravitation] stumbled on some theorems relating volume and surface [Lützen might have added *line*, as well] integrals. Therefore, the history of Green's, Gauss's, and Stokes's theorems are full of independent discoveries."[6] Gauss probably was the first to arrive at the theorem, but as was his style he did not publish it. What Gauss *didn't* publish would have made the reputations of ten mathematicians.

In 1828 Green *did* publish the theorem, in the privately printed "Essay on the Application of Mathematical Analysis to the Theories of Electricity and

Magnetism." By the very fact that it did not appear in an archival journal, Green's "Essay" quickly became rare, almost extinct.[7] After Green was admitted to Cambridge University in 1833 as a formal student, and his talent was appreciated by the established academics there, his "Essay" became a document that everybody had heard about but few had seen. In 1839 Green became a Fellow at Cambridge himself, but his career ended abruptly just two years later with his early death from either alcoholism or lung disease. Then, in 1845 and just after his graduation from Cambridge, William Thomson (later Lord Kelvin) asked his mathematics tutor about Green's "Essay" and was promptly given three copies of it.

Clearly Thomson knew of Green's theorem in 1845, but in a postscript to a letter (dated July 2, 1850) that he wrote to an academic friend at Cambridge, George Stokes (1819–1903), Thomson mentioned the theorem but neither gave a proof of it nor mentioned Green's authorship. In February 1854 Stokes made the proof of the theorem an examination question at Cambridge—on a test, it is amusing to note, taken by a youthful James Clerk Maxwell. Maxwell (1831–79) later developed the mathematical theory of electromagnetics in his 1873 masterpiece *Electricity and Magnetism* where, in a footnote to article 24, he attributes the theorem to Stokes. So, today, the theorem is often called— you guessed it—Stokes' theorem.

So, just whose theorem is it? If first publication is the criterion for deciding, then the answer has to be Green and he, in fact, seems to be the overwhelming choice by textbook writers, too. But what of Cauchy, you may ask? As best I have been able to untangle this whole business, the fundamental idea of Green's theorem *is* in Cauchy's 1814 memoir. But Cauchy's use of it is embedded in a paper whose central subject is complex integration, while Green made the theorem itself an explicit issue. Cauchy did, eventually, use the approach I am going to show you to prove the first integral theorem but, as I mentioned before, only after Green's "Essay" had been published.

To *really* add some spice to all this, recall that I mentioned earlier that Maxwell attributed what I am calling Green's theorem to Stokes in his *Electricity and Magnetism*. And yet, in article 96 of that book Maxwell *does* write of a Green's theorem—and he cites the "Essay." But there he is writing of the integral theorem that relates a double (surface) integral not to the single (line) integral around the edge of an open surface, but rather to a triple (volume) integral over the space enclosed by a closed surface. This is, today, usually called Gauss' theorem. And finally, I should tell you that, in Russia, Green's theorem goes by the name of Ostrogradski's theorem, after Mikhail Ostrogradski (1801–62), a Parisian acquaintance of Cauchy's in the 1820s who independently discovered it in 1831. Confused? Don't worry about it—the important thing is to know the *theorems*, not their names. With all that said, let's prove the theorem.

I am going to start by making the very simplifying assumption that R is a rectangular patch oriented parallel to the x and y axes as shown in figure 7.4, and that its boundary is $C = C_1 + C_2 + C_3 + C_4$—which simply means the boundary consists of four sides. Then, at the end, I'll try to convince you that the proof is actually far more general than this initial assumption may appear. So let us start by first considering the $\int_R\int - (\partial P/\partial y)dxdy$ term of the right-hand side of Green's theorem. We have

$$-\iint_R \frac{\partial P}{\partial y}\ dx\ dy = -\int_{x_0}^{x_1}\left\{\int_{y_0}^{y_1}\frac{\partial P}{\partial y}dy\right\}\ dx$$

$$= -\int_{x_0}^{x_1}\{P(x,y_1)-P(x,y_0)\}\ dx = \int_{x_0}^{x_1}P(x,y_0)\ dx + \int_{x_1}^{x_0}P(x,y_1)\ dx$$

$$= \int_{C_1}P(x,y)dx + \int_{C_3}P(x,y)dx.$$

Notice carefully that in the last two integrals I have dropped the subscripts on y_0 and y_1 that I did include in the previous integrals. I can do that because

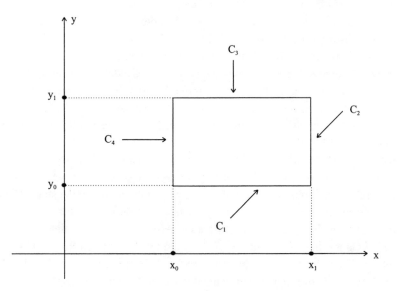

Figure 7.4. The prototype region for proving Green's theorem.

the subscripts were there to distinguish between integration along the lower edge (y_o) or along the upper edge (y_1) of the rectangle's boundary, and that job is now done in the last two integrals by writing C_1 (lower edge) and C_3 (upper edge) beneath the appropriate integral. Notice, too, that writing $\int_{y_o}^{y_1}(\partial P/\partial y)dy$ = $P(x,y_1) - P(x,y_o)$ makes the implicit assumption that there is no discontinuity in $\partial P/\partial y$, i.e., that the derivative is continuous. This is, in fact, nothing less than the fundamental theorem of integral calculus.

Similar integrals with respect to x can be written for the other two edges (C_2 and C_4) as well and, because they are vertical, along them we know $dx = 0$. Those two integrals, then, must vanish and we can formally add them to the C_1 and C_3 integrals without changing anything. So

$$-\iint_R \frac{\partial P}{\partial y} dx\, dy = \int_{C_1} P(x,y)dx + \int_{C_3} P(x,y)dx + \int_{C_2} P(x,y)dx + \int_{C_4} P(x,y)dx$$

$$= \int_C P(x,y)dx.$$

If you repeat the above for the $\int_R\int(\partial Q/\partial x)dxdy$ term in Green's theorem, and observe that $dy = 0$ along the horizontal edges C_1 and C_3, you can easily show that

$$\iint_R \frac{\partial Q}{\partial x} dx\, dy = \oint_C Q(x,y)\, dy.$$

And that completes the proof of Green's theorem for our nicely oriented rectangle. In fact, however, the above proof extends easily to other much more complicated shapes for R.

In figure 7.5, for example, I have shown a semicircular disk that can be built up out of many very thin rectangles—the thinner they each are the more of them there are, but that's okay; make them each as thin as the finest onion-skin paper—that very closely approximates the half-disk. If the boundary edge of the half-disk is denoted by C, and the boundary edges of the rectangles are denoted by $C_1, C_2, C_3 \ldots$, then obviously

$$\oint_C (Pdx + Qdy) = \int_{C_1} (Pdx + Qdy) + \int_{C_2} (Pdx + Qdy)$$

$$+ \int_{C_3} (Pdx + Qdy) + \cdots,$$

because those edges of the individual rectangular boundaries that are parallel to the x-axis are traversed twice, once in each sense (CW and CCW), and so their contributions to the various integrals on the right-hand side cancel

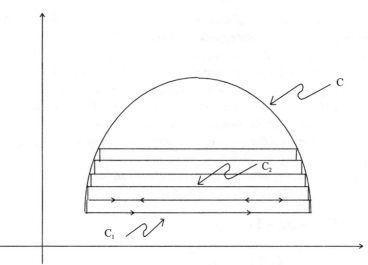

Figure 7.5. Approximating a simply connected region with arbitrarily many rectangles.

(you'll see this trick again before the chapter is over). The only exception is the very bottom horizontal edge, C_1. Only the integrations along the individual rectangular edges that are vertical avoid cancellation. If we make the rectangles really thin, the union of those vertical edges *is* C. The only shapes for R that I am going to use in this chapter are just rectangles, half-disks, and disks, and so what we have is now enough. But, in fact, Green's theorem applies to far more complicated shapes than I can possibly draw.

7.6 CAUCHY'S SECOND INTEGRAL THEOREM

Now you are ready for the *really* wonderful result from Cauchy's 1814 memoir. What happens, Cauchy asked, if we integrate a complex function around a contour that contains one (or more) points where the function is *not* analytic? To be precise, suppose $f(z)$ is analytic everywhere in the region that is the inside of C and which includes, in particular, the point $z = z_0$. Then $f(z)/(z - z_0)$ is also analytic everywhere in that region except at $z = z_0$, where $f(z)/(z - z_0)$ blows up. z_0 is what is called a first-order *singularity* or a *simple pole*. The term "first-order" is self-explanatory, I think, and by extension $f(z)/(z - z_0)^2$ and $f(z)/(z - z_0)^3$ have second- and third-order singularities at $z = z_0$. You can even run into infinite-order singularities, e.g., the power series expansion of the exponential shows that the function $e^{1/z}$ has one at $z = 0$.

Calling $z = z_o$ a pole, however, probably needs some explanation. The story I heard while an undergraduate electrical engineering student goes like this: Imagine a three-dimensional plot, with the real (x) and imaginary (y) axes using up two dimensions to form the complex plane, and $|f(z)|$ plotted on the third axis. The plot of $|f(z)|$ will form a surface in the space above the complex plane, undulating up and down as z varies in the plane below, much like a circus tent does above the performing arena. And the tent pokes upward particularly high at the location of the tent poles. Who says technical people aren't poets?[8]

Cauchy's question, then, was:

$$\oint_C \frac{f(z)}{z - z_0} \, dz = ?$$

Cauchy's answer to this question is his second integral theorem and, in my opinion, it is one of the most beautiful, profound, indeed mysterious results in all of mathematics. It is also easy to derive—but then, everything is easy when you see how someone of genius did it first!

In figure 7.6 I have drawn C and, in its interior, the point $z = z_o$. In addition, C^* is a circle centered on $z = z_o$, with a radius ρ that is sufficiently small that

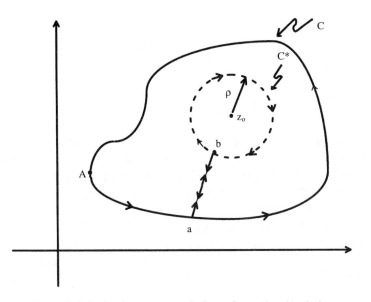

Figure 7.6. A simple contour enclosing a first-order singularity, connected to an inner circle with a cross-cut.

C^* is always inside of C. Now, imagine that we start our journey around C at some point A and move counterclockwise until we reach point a, whereupon we travel inward to point b on C^*. Once on C^* we travel clockwise (i.e., in the negative sense) until we return to point b. Then, we travel outward back to point a, retracing our original inward path, and continue the original inter- rupted trip around C until we return to our starting place, point A. Now here's the first really important thing about what we've done—*this path has always kept the annular region between* C *and* C* *to our left,* i.e., this path is the edge of a region in which the point $z = z_o$ is always on the outside. In that annular region, from which $z = z_o$ has by construction been excluded, $f(z)/(z - z_o)$ is analytic everywhere. Thus, by Cauchy's first integral theorem, since $z = z_o$ is outside of C then

$$\oint_{C,ab,-C^*,ba} \frac{f(z)}{z - z_0}\, dz = 0.$$

Now, here's the second important observation about this integration path. The two trips along the ab connection between C and C^* are in opposite directions, and so their individual contributions to the overall integral cancel—this two-way connection path between C and C^* is called a *cross-cut* by mathematicians. Thus, we can simplify the last statement I wrote to

$$\oint_{C,-C^*} \frac{f(z)}{z - z_0}\, dz = 0$$

or, in a form perhaps more revealing,

$$\oint_{C} \frac{f(z)}{z - z_0}\, dz - \oint_{C^*} \frac{f(z)}{z - z_0}\, dz = 0.$$

The reason for the minus sign in front of the C^* contour integral is that the actual trip around C^* was done in the negative sense (which is why I previ- ously wrote $-C^*$), but in the last expression the two integrals are both in the positive sense. That is, I've simply moved the minus sign from $-C^*$ to the front of the integral itself.

Let me remind you here that C is an arbitrary simple (non-self-intersecting) curve enclosing $z = z_o$, while C^* is a circle of radius ρ centered on $z = z_o$. Thus, on C^* we can write $z = z_o + \rho e^{i\theta}$ and so $dz = i\rho e^{i\theta}d\theta$ and therefore, as θ varies from 0 to 2π radians on one trip around C^*,

$$\oint_{C} \frac{f(z)}{z - z_0}\, dz = \int_{0}^{2\pi} \frac{f(z_0 + \rho e^{i\theta})}{\rho e^{i\theta}} i\rho e^{i\theta} d\theta = i \int_{0}^{2\pi} f(z_0 + \rho e^{i\theta})d\theta.$$

If the integral on the far left, i.e., the integral that is the original target of our interest, is to have a value, then whatever it is must be independent of the value of ρ. After all, the integral on the left has no ρ in it! So the value of the integral on the far right must be independent of ρ, too, and we can use any value we want—we will use one that's convenient. In fact, let us use a very small ρ. Indeed, we can pick ρ so small as to make the difference between $f(z)$ and $f(z_o)$, for all z on C^*, as small as we like. This is so because since $f(z)$ is analytic then it has a derivative everywhere inside of C (including at $z = z_o$) and so it is certainly continuous. Thus, as $\rho \rightarrow 0$ we can argue that $f(z) = f(z_o)$ all along C^* and so it is legal to pull the constant $f(z_o)$ out of the integral and write our answer: if $z = z_o$ is inside C then

$$\oint_C \frac{f(z)}{z - z_0} \, dz = if(z_0) \int_0^{2\pi} d\theta = 2\pi if(z_0).$$

Alternatively, we can write this as

$$f(z_0) = \frac{1}{2\pi i} \oint_C \frac{f(z)}{z - z_0} \, dz,$$

a form that tells us that the value of the analytic function $f(z)$ at the arbitrary interior point $z = z_o$ is completely determined by just the values of $f(z)$ along the boundary edge C. This intimate connection of the values of $f(z)$ inside C to the values of $f(z)$ on C is another illustration of the very special nature of analytic complex-valued functions, compared to just any old function written down at random. In fact, it is the global reach of an analytic function, from one part of the complex plane to another part, that is behind the property of analytic continuation that I discussed back in section 6.4. Now, let me show you what we can *do* with this result, called Cauchy's second integral theorem.

Let us evaluate the contour integral

$$\int_C \frac{e^{iaz}}{b^2 + z^2} \, dz,$$

where C is the contour shown in figure 7.7, with a and b positive constants, for the limiting case as $R \rightarrow \infty$. You'll see that this leads to an interesting result. Along the real axis part of C we have $z = x(dz = dx)$, and along the semicircular arc we have $z = Re^{i\theta}(dz = iRe^{i\theta}d\theta)$ where $\theta = 0$ at $x = R$ and $\theta = \pi$ radians at $x = -R$. So

$$\oint_C \frac{e^{iaz}}{b^2 + z^2} \, dz = \int_{-R}^{R} \frac{e^{iax}}{b^2 + x^2} \, dx + \int_0^{\pi} \frac{e^{ia(Re^{i\theta})}}{b^2 + R^2 e^{i2\theta}} i Re^{i\theta} \, d\theta.$$

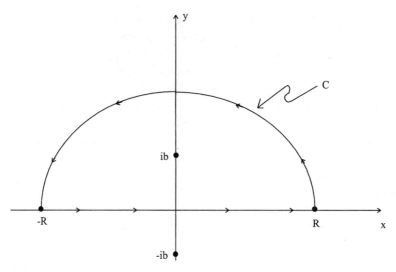

Figure 7.7. Another simple contour enclosing a single first-order singularity.

The integrand of the original contour integral, on the left-hand side, can be written as

$$\frac{e^{iaz}}{b^2 + z^2} = \frac{e^{iaz}}{(z+ib)(z-ib)} = \frac{e^{iaz}}{i2b}\left[\frac{1}{z-ib} - \frac{1}{z+ib}\right],$$

and so we have

$$\frac{1}{2ib}\left[\oint_C \frac{e^{iaz}}{z-ib}\,dz - \oint_C \frac{e^{iaz}}{z+ib}\,dz\right] = \int_{-R}^{R} \frac{e^{iax}}{b^2 + x^2}\,dx + \int_0^\pi \frac{e^{ia(Re^{i\theta})}}{b^2 + R^2 e^{i2\theta}}\,iRe^{i\theta}\,d\theta.$$

Since the integrand of the second contour integral on the left-hand side is analytic everywhere *inside* of *C*—that integrand does have a singularity, yes, but it's at $z = -ib$, which is *outside* of *C*, as shown in figure 7.7—then we know from Cauchy's first integral theorem that the contour integral is zero. Thus,

$$\frac{1}{i2b}\oint_C \frac{e^{iaz}}{z-ib}\,dx = \int_{-R}^{R} \frac{e^{iax}}{b^2 + x^2}\,dx + \int_0^\pi \frac{e^{ia(Re^{i\theta})}}{b^2 + R^2 e^{i2\theta}}\,iRe^{i\theta}\,d\theta.$$

And, once $R > b$ (remember, I'm eventually going to let $R \to \infty$) then the singularity for the remaining contour integral integrand is *inside C*, at $z = ib$. This integrand looks exactly like $f(z)/(z - z_o)$, with $f(z) = e^{iaz}$ and, of course, $z_o = ib$. Cauchy's second integral theorem tells us that, if $R > b$, the contour integral is equal to $2\pi i f(z_o)$, and so the left-hand side of the last equation is equal to

$$\frac{1}{i2b} 2\pi i \; e^{ia(ib)} = \frac{\pi}{b} e^{-ab}.$$

That is,

$$\int_{-R}^{R} \frac{e^{iax}}{b^2 + x^2}\, dx + \int_{0}^{\pi} \frac{e^{ia(Re^{i\theta})}}{b^2 + R^2 e^{i2\theta}}\; iRe^{i\theta}\; d\theta = \frac{\pi}{b} e^{-ab}, \quad R > b.$$

Now, if we at last let $R \to \infty$ then, using the same sort of argument given in box 7.1, *you* can show that the second integral on the left vanishes. Thus,

$$\int_{-\infty}^{\infty} \frac{e^{iax}}{b^2 + x^2}\, dx = \frac{\pi}{b} e^{-ab} = \int_{-\infty}^{\infty} \frac{\cos(ax)}{b^2 + x^2}\, dx + i \int_{-\infty}^{\infty} \frac{\sin(ax)}{b^2 + x^2}\, dx.$$

Equating real and imaginary parts we arrive at

$$\int_{-\infty}^{\infty} \frac{\sin(ax)}{b^2 + x^2}\, dx = 0,$$

which is no surprise since the integrand is an odd function of x, and also, for $a,b > 0$,

$$\int_{-\infty}^{\infty} \frac{\cos(ax)}{b^2 + x^2}\, dx = \frac{\pi}{b} e^{-ab}$$

which is, I think, *quite* a surprise! It had, in fact, been discovered by different means in 1810 by Laplace. For the special case of $a = b = 1$ this reduces to

$$\int_{-\infty}^{\infty} \frac{\cos(x)}{1 + x^2}\, dx = \frac{\pi}{e},$$

the integral I tantalized you with in the opening section to this chapter. Now, let's do a harder integral.

In his 1814 memoir Cauchy derived many interesting formulas to demonstrate the power of contour integration. One of the more famous of these calculations is the evaluation of

$$\int_{0}^{\infty} \frac{x^{2m}}{1 + x^{2n}}\, dx,$$

where m and n are both non-negative integers, with $n > m$. The answer is both surprising and beautiful. What I will show you now is how to use Cauchy's second integral theorem to do this, using the modern approach (you can find the details of Cauchy's original and somewhat awkward[9] analysis in Ett-linger's paper—see note 1).

I will begin with the contour integral

$$\oint_C \frac{z^{2m}}{1+z^{2n}}\,dz,$$

where C is a contour that contains some but not all the singularities of the integrand. I'll be more specific about the precise form of C in just a bit. Now, the singularities of the integrand are simply the solutions to the cyclotomic equation $1 + z^{2n} = 0$, which you know from the discussion of De Moivre's formula in Chapter Three are just the $2n$ $2n$th roots of $-1 = 1 \angle 180° = 1 \angle \pi$ radians. You know from that discussion that these roots are evenly spaced around the unit circle, with one of them the obvious $1 \angle \pi/2n$ radians. The others are spaced at intervals of $2\pi/2n$ radians, and so the $2n$ roots are at

$$z_k = 1 \angle \left(\frac{\pi}{2n} + k\frac{2\pi}{2n}\right) = 1 \angle \frac{2k+1}{2n}\pi$$
$$= e^{i\pi(2k+1)/2n},\ \ k = 0,\,1,\,2,\,\ldots,\,2n-1.$$

If this sounds familiar, it should—it is the same argument I used in section 3.3 when I factored the cyclotomic polynomial $z^{2n} - 1$.

As I argued there, it is clear by symmetry that half of these roots are in the upper half of the complex plane (those for $k = 0, 1, \ldots, n - 1$), and half are in the lower half of the complex plane (those for $k = n, n + 1, \ldots, 2n - 1$). The real axis, in fact, neatly separates these two halves—you can easily convince yourself that there are no roots *on* the real axis by showing that there is no integer k that gives a root angle of either zero or π radians. Now, let us pick the integration contour C to be that of figure 7.7, and so C will enclose only half of the singularities of the integrand, with the other half (in the lower part of the complex plane) excluded. You might think we aren't using all the information contained in the integrand by doing this, but don't forget the roots of $1 + z^{2n} = 0$ are conjugate *pairs*, i.e., if you know where the upper half-plane poles are then you automatically know where the lower half-plane poles are, too. Thus,

$$\oint_C \frac{z^{2m}}{1+z^{2n}}\,dz = \int_{-R}^{R} \frac{x^{2m}}{1+x^{2n}}\,dx + \int_{C_R} \frac{z^{2m}}{1+z^{2n}}\,dz,$$

where C_R is the semicircular arc. On the arc $z = Re^{i\theta}$ and so

214

$$\int_{C_R} \frac{z^{2m}}{1+z^{2n}} \, dz = \int_0^\pi \frac{R^{2m} e^{i2m\theta}}{1+R^{2n} e^{i2n\theta}} \, iRe^{i\theta} \, d\theta.$$

Since $n > m$, then R^{2n} is at least one degree greater than R^{2m+1} and so, as $R \to \infty$, the magnitude of the C_R integral will go to zero at least as fast as $1/R$. Thus,

$$\oint_C \frac{z^{2m}}{1+z^{2n}} \, dz = \int_{-\infty}^\infty \frac{x^{2m}}{1+x^{2n}} \, dx = 2\int_0^\infty \frac{x^{2m}}{1+x^{2n}} \, dx.$$

So, if we can do the contour integral on the left, then we will have done the real integral on the right, and we can do the contour integral using Cauchy's second integral theorem. Here's how.

Since the integrand denominator can be written as

$$1 + z^{2n} = \prod_{k=0}^{2n-1} (z - z_k),$$

in which each factor in the product is of the first degree, then the singularities of the integrand are all simple poles. Now, I claim that the integrand can be written as a partial fraction expansion, i.e., as

$$\frac{z^{2m}}{1+z^{2n}} = \frac{N_0}{z - z_0} + \frac{N_1}{z - z_1} + \frac{N_2}{z - z_2} + \cdots + \frac{N_{2n-1}}{z - z_{2n-1}},$$

where the N's are constants. I will prove this by the direct approach of actually calculating them.

But before doing that, you should understand *why* I'm going to do it. Suppose, in fact, that I already knew the N's. Then I could write

$$\oint_C \frac{z^{2m}}{1+z^{2n}} \, dz = \left\{ \oint_C \frac{N_0}{z - z_0} \, dz + \cdots + \oint_C \frac{N_{n-1}}{z - z_{n-1}} \right\}$$
$$+ \left\{ \oint_C \frac{N_n}{z - z_n} \, dz + \cdots + \oint_C \frac{N_{2n-1}}{z - z_{2n-1}} \right\}.$$

Since z_o through z_{n-1} are inside C, and since z_n through z_{2n-1} are outside C, then Cauchy's first integral theorem says that all of the integrals in the rightmost braces vanish, and Cauchy's second integral theorem says that the integrals in the first set of braces are just $2\pi i N_o, 2\pi i N_1, \ldots, 2\pi i N_{n-1}$, i.e., that

$$\oint_C \frac{z^{2m}}{1+z^{2n}} \, dz = 2\pi i \sum_{k=0}^{n-1} N_k = 2 \int_0^\infty \frac{x^{2m}}{1+x^{2n}} \, dx.$$

This follows from the fact that

$$\oint_C \frac{1}{z - z_0} \, dz = 2\pi i$$

if z_0 is inside C, with the interpretation in Cauchy's second integral theorem of $f(z) = 1$ and so, in particular, $f(z_0) = 1$. The answer to our problem is then simply

$$\int_0^\infty \frac{x^{2m}}{1+x^{2n}} \, dx = \pi i \sum_{k=0}^{n-1} N_k.$$

That's why the N's are important to us. Here is how to calculate them.

Let us target one specific N, say N_p, where $0 \leq p \leq n - 1$. Multiplying through the partial fraction expansion for the integrand by $(z - z_p)$, we get

$$\frac{(z-z_p)z^{2m}}{1+z^{2n}} = \frac{(z-z_p)N_0}{z-z_0} + \frac{(z-z_p)N_1}{z-z_1} + \cdots + N_p + \frac{(z-z_p)N_{p+1}}{z-z_{p+1}} + \cdots.$$

We can solve for N_p by letting $z \to z_p$, a process that makes every term on the right vanish, except for the N_p term that has no $(z - z_p)$ factor to carry it to zero. That is,

$$N_p = \lim_{z \to z_p} \frac{(z-z_p)z^{2m}}{1+z^{2n}} = \lim_{z \to z_p} \frac{z^{2m+1} - z_p z^{2m}}{1+z^{2n}}.$$

Since p is arbitrary I have, in fact, thus found all of the N's. This limit has the indeterminate form 0/0, and so we can use L'Hôpital's rule to evaluate it. This rule, proved in every freshman calculus textbook, says that if $\lim_{x \to x_0} f(x)/g(x) = 0/0$ then the limit is $\lim_{x \to x_0} f'(x)/g'(x)$ unless this also gives an indeterminate result (which means that one applies the rule once again). The rule was first published in the 1696 textbook *Analyse des Infiniment Petits,* authored by the Marquis Guillaume F. A. de L'Hôpital (1661–1704). It was originally discovered, however, by Euler's mentor John Bernoulli, who had communicated the rule to L'Hôpital in a letter.

Continuing, then,

$$N_p = \lim_{z \to z_p} \frac{(2m+1)\, z^{2m} - 2mz^{2m-1}z_p}{2n\, z^{2n-1}} = \frac{z_p^{2(m-n)+1}}{2n}.$$

That is,

$$N_p = \frac{e^{i\pi\frac{2p+1}{2n}(2m+1-2n)}}{2n} = \frac{e^{i\pi\frac{(2p+1)(2m+1)}{2n}}e^{-i2\pi n}}{2n}$$

$$= -\frac{e^{i\pi\frac{(2p+1)(2m+1)}{2n}}}{2n}.$$

So,

$$\int_0^\infty \frac{x^{2m}}{1+x^{2n}}\,dx = -\frac{\pi i}{2n}\sum_{p=0}^{n-1} e^{i\pi\frac{(2p+1)(2m+1)}{2n}}.$$

The summation is a geometric series with common factor between adjacent terms of $e^{i\pi 2(2m+1)/2n}$. If you are careful in your algebra (remember the trick in section 5.2), and remember Euler's identity, you should be able to show that the sum is

$$-i\frac{1}{\sin\left(\frac{2m+1}{2n}\pi\right)}$$

Multiplying this by $-\pi i/2n$ we finally arrive at the result

$$\int_0^\infty \frac{x^{2m}}{1+x^{2n}}\,dx = \frac{\pi}{2n\sin\left(\frac{2m+1}{2n}\pi\right)}.$$

It is standard today to actually write $2m + 1 = \alpha$ and $2n = \beta$ so that the integral becomes

$$\int_0^\infty \frac{x^{\alpha-1}}{1+x^\beta}\,dx = \frac{\pi}{\beta\sin\left(\frac{\alpha}{\beta}\pi\right)},$$

a result that Euler had derived (by other means) in 1743. You'll recall that I used this result back in section 6.13 in deriving the reflection formula for Euler's gamma function. Notice that, for the special case of $\alpha = 1$ ($m = 0$) and $\beta = 2$ ($n = 1$), the integral reduces to

$$\int_0^\infty \frac{dx}{1+x^2} = \frac{\pi}{2\sin\left(\frac{\pi}{2}\right)} = \frac{\pi}{2},$$

217

which agrees with the well-known definite integral

$$\int_0^\infty \frac{dx}{1+x^2} = \tan^{-1}(x) \Big|_0^\infty = \tan^{-1}(\infty) - \tan^{-1}(0) = \frac{\pi}{2}$$

(e.g., look back at the discussion in section 6.7 of Count Fagnano) definite integral

By now you have certainly spotted the general idea behind Cauchy's evaluation of improper definite integrals of the form $\int_{-\infty}^\infty f(x)dx$. There are two steps, in fact, in doing such integrals by contour integration in the complex plane: first, the selection of the proper complex function, and second the selection of a contour that, after some limiting process, includes the entire finite real axis as the $z = x$ part of the integration path. Along the real axis the complex function must reduce to $f(x)$. There are a number of complications that can occur in this process that I have not discussed, and that is what I meant in the introduction to this chapter when I asked the question "Where to stop?" Well, we're getting close to stopping, but not quite yet. There is one, last calculation I want to show you, both illustrating how to do an entirely different kind of definite, *proper* integral and tidying up a point I left undone in chapter 5.

7.7 KEPLER'S THIRD LAW: THE FINAL CALCULATION

In chapter 5 I told you that

$$\int_0^{2\pi} \frac{d\theta}{\{1 + E\cos(\theta)\}^2} = \frac{2\pi}{(1-E^2)^{3/2}}, \quad 0 \le E < 1.$$

This integral occurs in the general derivation of Kepler's third law, which I did only for the special case of circular orbits, i.e., orbits with eccentricity $E = 0$. So, as my very last example of a specific contour integral, let me show you how to use complex functions to evaluate the above trigonometric integral as a function of E.

The limits of 0 to 2π radians on the integral are a strong hint of a closed-loop trip around some contour in the complex plane, but what contour and with what complex function? We can answer both of these questions at the same time by taking our cue from Euler's identity and writing $z = e^{i\theta}$. Then,

$$\cos(\theta) = \frac{e^{i\theta} + e^{-i\theta}}{2} = \frac{z + (1/z)}{2}$$

and

$$dz = i\, e^{i\theta}\, d\theta = iz\,d\theta.$$

Thus,

$$\int_0^{2\pi} \frac{d\theta}{\{1 + E\cos(\theta)\}^2} = \oint_C \frac{dz}{iz\left\{1 + E\dfrac{z + (1/z)}{2}\right\}^2},$$

where C is the contour $|z| = 1$, i.e., C is the boundary edge of the unit circle.
After some elementary algebra you can get this into the form

$$\frac{4}{iE^2}\oint_C \frac{z\,dz}{\left(z^2 + \dfrac{2}{E}z + 1\right)^2} = \frac{4}{iE^2}\oint_{|z|=1} \frac{z\,dz}{\left(z + \dfrac{1}{E} + \sqrt{\dfrac{1}{E^2} - 1}\right)^2\left(z + \dfrac{1}{E} - \sqrt{\dfrac{1}{E^2} - 1}\right)^2}.$$

That is, the integrand has two singularities, each second order, at

$$z = \frac{1}{E} \pm \sqrt{\frac{1}{E^2} - 1}.$$

Since we are interested specifically in closed, repeating (periodic) satellite orbits, we then know $0 \le E < 1$, and so both poles are on the negative real axis. There is, however, a crucial difference between these two poles: the one at $-1/E - \sqrt{1/E^2 - 1}$ is *outside* C while the one at $-1/E + \sqrt{1/E^2 - 1}$ is *inside* C. Thus, our integral has the form

$$\frac{4}{iE^2}\oint_C \frac{f(z)}{(z - z_0)^2}\,dz,$$

where $f(z)$ is analytic everywhere on and inside C, i.e.,

$$f(z) = \frac{z}{\left(z + \dfrac{1}{E} + \sqrt{\dfrac{1}{E^2} - 1}\right)^2} \quad \text{and} \quad z_0 = -\frac{1}{E} + \sqrt{\frac{1}{E^2} - 1}.$$

But we seem to have a problem! The integral is not of the form in Cauchy's second integral theorem because the denominator is not just $(z - z_0)$, but rather $(z - z_0)^2$. That is, we have a second-order pole, not a simple one. So, what do we do now? As it turns out, we really don't have to do very much more at all.

Writing Cauchy's second integral theorem out again, we have, for z_0 any arbitrary point inside C,

219

$$f(z_0) = \frac{1}{2\pi i} \oint_C \frac{f(z)}{z - z_0} \, dz$$

and so

$$f(z_0 + \Delta z) = \frac{1}{2\pi i} \oint_C \frac{f(z)}{z - (z_0 + \Delta z)} \, dz = \frac{1}{2\pi i} \oint_C \frac{f(z)}{(z - z_0) - \Delta z} \, dz.$$

Then, plugging this into the definition of the derivative of $f(z)$,

$$f'(z_0) = \lim_{\Delta z \to 0} \frac{f(z_0 + \Delta z) - f(z_0)}{\Delta z} = \lim_{\Delta z \to 0} \frac{1}{2\pi i \Delta z} \oint_C \left\{ \frac{f(z)}{(z - z_0) - \Delta z} - \frac{f(z)}{z - z_0} \right\} dz$$

$$= \lim_{\Delta z \to 0} \frac{1}{2\pi i \Delta z} \oint_C f(z) \frac{(z - z_0) - (z - z_0) + \Delta z}{(z - z_0)^2 - \Delta z (z - z_0)} \, dz$$

$$= \frac{1}{2\pi i} \oint_C \frac{f(z)}{(z - z_0)^2} \, dz.$$

Or, at last, we have

$$\oint_C \frac{f(z)}{(z - z_0)^2} \, dz = 2\pi i \; f'(z_0).$$

Now our original integral looks just like the contour integral on the left of this equation. This trick for getting Cauchy's second integral theorem, derived for a simple pole, into a form involving a second-order singularity, can obviously be generalized for any order of singularity. Just keep differentiating. The general result is that

$$f^{(n)}(z_0) = \frac{n!}{2\pi i} \oint_C \frac{f(z)}{(z - z_0)^{n+1}} \, dz,$$

where $f^{(n)}(z_0)$ denotes the nth derivative of $f(z)$ evaluated at the integrand pole $z = z_0$ which, of course, is inside C. In his 1814 memoir, Cauchy considers only simple poles. For our problem we have

$$\frac{4}{iE^2} \cdot 2\pi i \; f'\left(-\frac{1}{E} + \sqrt{\frac{1}{E^2} - 1}\right) = \frac{8\pi}{E^2} f'\left(-\frac{1}{E} + \sqrt{\frac{1}{E^2} - 1}\right).$$

If you differentiate

$$f(z) = \frac{z}{\left(z + \dfrac{1}{E} + \sqrt{\dfrac{1}{E^2} - 1}\right)^2},$$

and then evaluate the result at $z = z_o$, you will find that

$$\int_0^{2\pi} \frac{d\theta}{\{1 + E\cos(\theta)\}^2} = \frac{8\pi}{E^2} \cdot \frac{E^2}{4(1-E^2)^{3/2}} = \frac{2\pi}{(1-E^2)^{3/2}},$$

the answer I gave you in chapter 5.

7.8 Epilog: What Came Next

NON SEQUITUR by Wiley

Contour integration before Cauchy?

After his 1814 memoir, Cauchy spent the next three and a half decades slowly laying the foundations of complex function theory. He didn't do everything, of course, but nearly so. Even in the pioneering work of others Cauchy was influential. This is illustrated by the important contribution of Pierre Alphonse Laurent (1813–54). Like Cauchy, Laurent was originally trained as a civil engineer and he worked for some years on hydraulic construction projects as an officer in the French military engineering corps. Before Laurent's second career as a mathematician began, Cauchy was unaware of the power series expansion form for analytic functions. Then, in 1843, Laurent made great progress in this area ("Laurent series" is a topic in every undergraduate textbook on complex variables), but his work was known during his lifetime only because Cauchy wrote about it to the French Academy, and encouraged publication. For some reason that was not to happen, however, until 1863, long after the deaths of both men.

Other nineteenth-century work in analytic function theory that was destined to have enormous impact on twentieth-century technology includes *conformal mapping,* and the theory of *system stability*. I will say just a few, brief words about both, but each could easily have a complete book all to itself—and, indeed, each has had *many* books. First, conformal mapping. This is the gen-

221

eral method of taking a complicated shape in the complex plane and then trying to discover a transformation equation that *maps* (or redraws) the boundary of that shape into some much simpler shape, e.g., a circle or a rectangle. One reason this is important is because Laplace's equation is often easy to solve for such simple shapes, where u and v of the complex function $f(z) = u + iv$ are defined along the boundary edge. (It can be shown, in fact, that this also defines u and v at all points inside the shape, as well.) Then, applying the conformal transformation equation, one arrives at the solution to Laplace's equation for u and v given along and inside the boundary of the original complicated shape. The names of the Germans Herman Amandus Schwarz (1842–1921) and Elwin Bruno Christoffel (1829–1900)—the latter was one of Dirichlet's students—are associated with this technique. It is treated in detail in all good undergraduate textbooks on complex variables.[10]

Stability theory, as studied by physicists and engineers, inevitably leads to time functions of the form $e^{(\sigma + i\omega)t}$, where σ and ω are both real. This form blows up as $t \to \infty$ if $\sigma > 0$, and so $\sigma \leq 0$ is the condition for a finite (or stable) behavior for the system being analyzed. Now, it often turns out to be the case that $s = \sigma + i\omega$ is a complex root of some equation $f(s) = 0$. If $f(s)$ is a polynomial, then $f(s)$ is analytic. Often, the question is not what are the specific value(s) of σ, but rather simply whether all the values of $\sigma \leq 0$. That condition insures the stability of the system. The problem of determining if all the solutions to $f(s) = 0$ have nonpositive real parts was posed in 1868 by Maxwell (who became interested in stability problems while studying the dynamics of Saturn's rings in the middle 1850s). It was solved by algebraic means in 1877 by Maxwell's rival at Cambridge, Edward John Routh (1831–1907). Later, in 1895, the German mathematician Adolf Hurwitz (1859–1919) solved the problem using complex function ideas. Today, all electrical engineers are taught the Routh-Hurwitz method for polynomial system stability.

Even before Cauchy's 1814 memoir, the early part of the nineteenth century saw the momentous work of Jean Baptiste Joseph Fourier (1768–1830), in 1807, on Fourier series expansions of periodic signals, and Fourier integrals for nonperiodic signals. In more advanced books it is shown that the Fourier integral is really a contour integral in the complex plane. That work, and its close relation, the Laplace transform, are used by physicists and electrical engineers all around the world, every day. I am not going to bother even to try to define these new terms, however, because this is, really and truly, where the book ends.

I simply want to conclude by telling you that by 1850 complex function theory was highly advanced—to the point that when Maxwell wrote *Electric-*

ity and Magnetism in 1873 he could open Article 183 with the following terse passage, with no further explanation: "Two quantities α and β are said to be conjugate functions of x and y, if $\alpha + \sqrt{-1}\,\beta$ is a function of $x + \sqrt{-1}\,y$. It follows from this definition that

$$\frac{d\alpha}{dx} = \frac{d\beta}{dy} \text{ and } \frac{d\alpha}{dy} + \frac{d\beta}{dx} = 0."$$

These are, of course, just the Cauchy-Riemann equations. But they certainly do not follow from Maxwell's definition of a complex function unless he could safely assume that all his readers understood the requirement for the derivative to be independent of how Δx and y vanish. Apparently, by 1873, just two decades after Riemann had first made that argument in his doctoral dissertation, Maxwell *could* make such an assumption.

In his 1851 doctoral dissertation Riemann, who succeeded Dirichlet as Professor of Mathematics at Göttingen upon Dirichlet's early death in 1859, set off yet another explosion of growth, following Cauchy's, with his discovery of the connection between multivalued complex functions (e.g., the log function) and topology. We saw just a hint of how topology comes into play with complex functions in the discussion in section 7.5 on simply and multiply connected regions. It was just such an advance in theory that was needed to do Riemann's integral that occurs in the derivation of the functional equation for the zeta function (presented in section 6.13). And, by the way, you can now appreciate why s is traditionally used as the complex variable in that integral. To evaluate the integral using contour methods we would replace the real dummy variable x with the complex dummy variable z, i.e., there are two complex variables involved in this integral, and so we need two symbols (s and z).[11] But all that is for another book.

In the November 1942 issue of the science fiction pulp magazine *Super Science Stories,* Isaac Asimov published a story called "The Imaginary."[12] This fanciful tale concerns an alien psychologist who has discovered how to model mental processes by pure mathematics, mathematics involving $\sqrt{-1}$. One of his colleagues is critical of this work, however, claiming that such equations couldn't possibly have any meaning. To this, a supporter of the new theory replies:

Meaning! Listen to the mathematician talk. Great space, man, what have mathematics to do with meaning? Mathematics is a tool and as long as it can be manipulated to give proper answers and to make correct predictions, actual meaning has no significance.

To that another doubter says:

> "I guess so. I guess so. But using imaginary quantities in psychological equations stretches my faith in science just a little bit. Square root of minus one!" He shuddered.

A second supporter ends the discussion with

> Do you suppose he cares how many imaginaries there are in the intermediate steps if they all square out into minus one in the final solution. All he's interested in is that they give him the proper sign in the answer. . . . As for its physical significance, what matter? Mathematics is only a tool, anyway.

I have mixed feelings about this fictional discussion. Mathematics *is* a tool for practical engineers and applied scientists, of course, but I suspect that even the purest of mathematicians finds a mental image of the symbols being manipulated helpful. It was the great Gauss himself, after all, who finally made the case for the geometric interpretation of $\sqrt{-1}$. I'll leave you with this metaphysical question, for you to arrive at your own personal conclusion. For me, however, the connection of rotation in the complex plane and multiplication by $\sqrt{-1}$ is unbreakable.

The intimate connection of $\sqrt{-1}$ to physical reality is still not always appreciated, however, even by educated people who claim to know quite a bit about mathematics. Consider, for example, these words—words that sound very much like Auden's couplet in section 1.2—from "celebrity intellectual" Marilyn vos Savant:

> The square root of $+1$ is a real number because $+1 \times +1 = +1$; however, the square root of -1 is imaginary because -1 times -1 would also equal $+1$, instead of -1. This appears to be a contradiction. Yet it is accepted, and imaginary numbers are used routinely. But how can we justify using them to *prove* a contradiction?[13]

Vos Savant's words reveal a curious lack of sophistication in her understanding of the complex plane. The real numbers, of which she apparently has no fear, are no more (or less) trustworthy than are the complex ones.

Well, perhaps vos Savant simply had a bad few weeks when she wrote (perhaps she should have taken a little more time to write her book). So, with a rueful chuckle at how even those blessed with a high IQ can write nonsense, this then *is* the end of this book. We have, indeed, covered a lot of ground between Heron and his "impossible" pyramid, Cardan's cubic formula, and Cauchy's contour integrals. And it is all the result of the question that opened this book—just what *does* $\sqrt{-1}$ mean? As I hope this book has convinced you, it means a great deal. In May 1799 Thomas Jefferson received a letter from a correspondent who had just finished a study of Euclid. The writer wished to learn of what other mathematics an educated man about to enter the nineteenth century should be knowledgeable. In a long and detailed letter dated June 19, 1799 Jefferson replied from his home Monticello, writing that trigonometry "is most valuable to every man," and also that "the extraction of square and cube roots, algebra as far as the quadratic equation, the use of logarithms are of value . . . but all beyond these is but a luxury; a delicious luxury indeed, but not to be indulged in by one who is to have a profession to follow for his subsistence."[14] Jefferson wrote these words just two years after

A SPLASH FROM THE COMPLEX PLANE

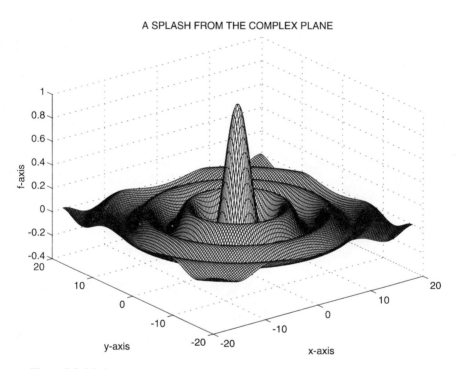

Figure 7.8. Modern computers have made the visualization of functions of complex variables easy to achieve. For example, the above image is the three-dimensional surface defined by $f = \sin(|z|)/|z|$. It is sometimes called the "splash" or the "rock-in-the-pond" function, for the obvious reason. It was created on my IBM ThinkPad 365ED laptop computer (using an 80486 processor chip @ 75 MHz) while running the powerful mathematical programming language MATLAB. To construct the image required 295,000 floating-point arithmetic operations, which the computer performed in less than two seconds. Do you think Euler, Cauchy, and Riemann would have had fun with a gadget that can do things like this?

225

Wessel had written his great paper on $\sqrt{-1}$, and surely Jefferson would have thought $\sqrt{-1}$ a "luxury."

One can only wonder at what Jefferson would have thought of our modern times (what would he have thought of figure 7.8?), in which so many use $\sqrt{-1}$ in the pursuit of their profession and subsistence. I think he would have thought any such claim of things to come the result of a fevered brain, and would have called such a prophecy "an imaginary tale." But now, as you close this book, you can appreciate the ironic truth in the fact that there is nothing at all imaginary about $\sqrt{-1}$.

The Fundamental Theorem of Algebra

THE THEOREM that is the subject of this appendix is easy to state, plausible, and *not* so easy to prove. Indeed, it is difficult enough that the problem was central to Gauss' 1799 doctoral dissertation. Gauss' 1799 proof is considered to be unacceptable by modern standards because he made an assumption in his proof that he thought "obvious," but which was not actually established until 1920. Gauss obviously had some reservations himself, as he later returned to the problem, and by the end of his life had offered three other proofs.[1] What the theorem says is that every nth-degree polynomial equation

$$f(z) = a_n z^n + a_{n-1} z^{n-1} + \cdots + a_1 z + a_0 = 0,$$

where the a's are arbitrary numbers—perhaps even complex, as was first pointed out by Argand, *not* Gauss—and n is any positive integer, always has exactly n roots in the complex number system. An important detail is that if there are identical roots of multiplicity m, then they count as m roots, not just one.[2]

Mathematicians long before Gauss had believed the theorem to be true. For example, Descartes stated in his *La Geometrie* (1637) "Every equation can have as many distinct roots (values of the unknown quantity) as the number of dimensions of the unknown quantity in the equation." Descartes' "proof" of the fundamental theorem, as an intuitive one, is easily grasped. It simply says that every root r of the polynomial equation $f(x) = 0$ must appear as the factor $(x - r)$ in the factorization of $f(x)$. And if $f(x)$ has degree n, then it will require n such factors (and hence n roots) to give the required x^n term. Descartes then simply illustrates this idea in *La Geometrie* with several specific examples, but offers no general proof.

Even earlier, his fellow countryman Albert Girard (1590–1632) wrote, in his *L'Invention Nouvelle en L'Algebra* (1629), "Every equation of algebra has as many solutions as the exponent of the highest term indicates." In particular, Girard discusses the roots of the equation $x^4 - 4x + 3 = 0$ and observes that, along with the double real root of 1, there are two additional complex roots, $-1 \pm i\sqrt{2}$, thus giving the required four roots. Before Gauss, some other famous mathematicians had tried their hands at actually proving the fundamental theorem, including d'Alembert, Euler, and Lagrange. All failed.

To summarize, the fundamental theorem is easy to understand and believe, but its proof was a tough nut to crack. We will take it on faith in this book. Not nearly so hard to prove, however, is another result that is of great value in working with complex numbers. It is the claim that when complex roots do appear as solutions to the polynomial equation $f(z) = 0$ with *real* coefficients, they appear as *conjugate pairs*. That is, if $z = x + iy$ is a root, then so is $\bar{z} = x - iy$. This was first shown by d'Alembert in 1746. To prove this assertion it will be helpful first to establish the preliminary result that the conjugate of the sum or product of two complex numbers, z_1 and z_2, is the sum or product of their conjugates. That is,

$$\overline{z_1 + z_2} = \bar{z}_1 + \bar{z}_2,$$
$$\overline{z_1 z_2} = \bar{z}_1\, \bar{z}_2.$$

The most direct way to show this is simply by direct algebraic substitution. That is, write $z_1 = a_1 + ib_1$ and $z_2 = a_2 + ib_2$, evaluate both sides of the claimed equalities, and observe that the results *are* equal.

Now, suppose that all the a's in the polynomial equation $f(z) = 0$ are real. From the very structure of $f(z)$, made up of just products and sums of complex numbers, we immediately have the result $f(\bar{z}) = \bar{f}(z)$ because of the two preliminary results established in the previous paragraph. Suppose next that z_1 is a solution to $f(z) = 0$. Then $\bar{z}$ is also a solution because $f(\bar{z}_1) = \bar{f}(z_1) = \bar{0} = 0$, i.e., like any real number, zero is its own conjugate. Do you see now why the a's must be real for the roots to appear as conjugate pairs? It is because when we form $\bar{f}(z)$ we generate the conjugates of the a's, too, as well as of z and its powers, and to keep them from mixing things up we need every $\bar{a} = a$, i.e., every a real. Thus, the complex roots of a polynomial equation of *any* degree *with real coefficients* (this proviso is important[3]) occur as complex conjugate pairs. In particular, the three roots to a cubic polynomial either must all be real, or there must be one real root and one conjugate pair. There cannot be three complex roots because then there would have to be one complex root without a conjugate mate.

To Be Read after Reading about Wessel's Work in Chapter Three

As a final comment on conjugates, notice that it is also true for any complex number $z = x + iy$ that

$$\overline{\sqrt[k]{x - iy}} = \sqrt[k]{\overline{x + iy}}.$$

That is, the conjugate of the *k*th root is the *k*th root of the conjugate. This is most easily shown using the polar form. Thus, we can write, with *n* any integer,

$$x + iy = \sqrt{x^2 + y^2}\ e^{i\{\tan^{-1}(y/x)+2\pi n\}}$$

and so

$$\sqrt[k]{x+iy} = (x^2 + y^2)^{1/2k}\ e^{i\{(1/k)\tan^{-1}(y/k)+2\pi n/k\}}.$$

Thus

$$\overline{\sqrt[k]{x+iy}} = (x^2 + y^2)^{1/2k}\ e^{-i\{(1/k)\tan^{-1}(y/k)+2\pi n/k\}}.$$

If you write out $\sqrt[k]{\overline{x+iy}}$ in polar form in the same way, then you'll find you get the same result.

For example, since $1 + \sqrt{-3}$ and $1 - \sqrt{-3}$ are conjugates, their square roots are also conjugates, and so Leibniz's famous expression (stated at the end of chapter 1) becomes obvious. Leibniz was so fascinated by this behavior, without really knowing *why* it is always so, that his unpublished papers contain many similar results. Of this the Caltech mathematician and historian of mathematics Eric Temple Bell wrote "He was . . . astonished by a . . . verification of the expression of a real [number] as a sum of conjugate complexes. The truly astonishing thing historically about Leibniz's performances with complex numbers is that less than three centuries ago [in 1940, when Bell first wrote these words] one of the greatest mathematicians in history should have thought that [this] outcome was more unexpected than is that of turning a tumbler upside down twice in succession."[4] With these arrogant words, of course, Bell is practicing Whig history, sneering at the ignorance of the past from the vantage point of the three hundred years of progress that came after Leibniz died.

The Complex Roots of a Transcendental Equation

IN THIS APPENDIX I will show you how complex numbers can be used to determine the nature of the roots of an equation *without* actually solving the equation. The example I will use is based upon one 1920s mathematician's interest in a purely technical issue that he read about in the contemporary physics literature. I'll stick here to the purely mathematical aspects of the problem, and let you look up the inspiration for it all, if you're interested.[1] The lesson to be learned from this example is not how to solve a particular equation for its particular roots. Rather, it is to see the sort of reasoning and arguments, based on complex numbers, that can lead to much information about the nature of an equation's roots without calculating the roots at all.

The equation to be studied is $f(z) = (1 + 2z)e^z - z = 0$. Since the power series expansion of e^z developed in chapter 6 has all powers of z present, then we have an equation of infinite degree and we can expect there to be an infinity of roots. First, let us establish that, in this infinity, there is not a single real root. To do this, suppose $z = x$ where x is real. Then,

$$e^x = \frac{x}{1 + 2x}$$

and figure B.1 shows sketches of the left- and right-hand sides of this equation. As shown in the figure, the curves for e^x and $x/(1 + 2x)$ have no intersections and so there is no real x. This may or may not seem "obvious" to you. Pierpont wrote of these curves, "we see at once [that they] do not meet in a real point," but *I* didn't see it in an instant and if you need a hint, too, perhaps the following will help.

The two branches of the hyperbola $x/(1 + 2x)$ (A and C in figure B.1) have the asymptotes shown in dashed lines. It is clear, I think, that the exponential curve (B) does not intersect A. The case of the exponential curve and branch C is just slightly less clear. Since e^x is less than $\frac{1}{2}$ for $x < -\ln(2) \approx -0.69$, then the only a priori possible interval for an intersection (i.e., a real root) would be in the interval $-\ln(2) < x < -0.5$. The slope of the exponential is e^x, and so at $x = -\ln(2)$ the slope is $\frac{1}{2}$. The slope of branch C is $1/(1 + 2x)^2$, and so at $x = -\ln(2)$ the slope is approximately $1/(1 - 2 \times 0.69)^2 = 1/(0.38)^2 > 1$. In fact, the slope of the exponential is always less than that of branch C, and

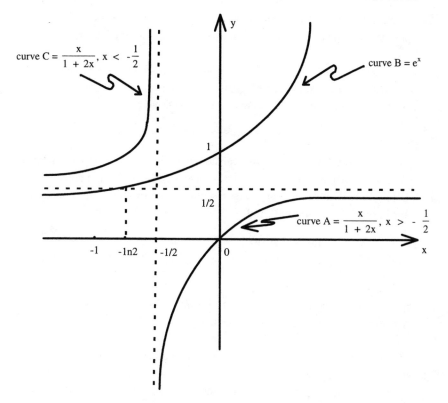

curve C = $\dfrac{x}{1 + 2x}$, x < $-\dfrac{1}{2}$

curve B = e^x

curve A = $\dfrac{x}{1 + 2x}$, x > $-\dfrac{1}{2}$

y

1

1/2

-1   -ln2   -1/2   0   x

Figure B.1. Geometric statement that $f(z) = (1 + 2z)e^z - z = 0$ has no real roots.

since C is above the exponential at $x = -\ln(2)$ then the two curves can never intersect.

Since there are no real roots, then all the roots must be complex. In fact, we can say something even stronger; *all* the roots are off the imaginary axis, i.e., there are no purely imaginary roots. To see this, suppose that there *are* imaginary roots $z = iy$, where y is real. Then,

$$(1 + 2iy)e^{iy} = iy,$$

and if we take the absolute value of both sides we have

$$|(1 + 2iy)e^{iy}| = |1 + 2iy| \, |e^{iy}| = |iy|.$$

But, as $|e^{iy}| = 1$ then this says that $1 + 4y^2 = y^2$ or $3y^2 = -1$. This is, of course, not possible for y real. This contradiction means that the original as-

231

sumption that $z = iy$ must be false, and so all of the roots of $f(z) = 0$ have nonzero real parts. Thus, *all* the roots are of the general form $z = x + iy$, $x \neq 0$.

But, once again, we can say even more. As I'll show you next, each of the roots has a conjugate mate, i.e., if z is a root then so is $\bar{z}$. Writing $z = x + iy$ and inserting into the original equation for $f(z)$, we have

$$[1 + 2(x + iy)]e^{x+iy} - (x + iy) = 0.$$

If you expand all of this out using the connection between complex exponentials and the trigonometric functions—Euler's identity from chapter 3—and collect the real and imaginary parts together (each of which must separately vanish, of course), you will find that

$$[(1 + 2x)\cos(y) - 2y\sin(y)]e^x - x = 0,$$
$$[(1 + 2x)\sin(y) + 2y\cos(y)]e^x - y = 0.$$

Now, if the replacements $x \to x$ and $y \to -y$ are made, which obviously represent a change from z to $\bar{z}$, then you find that these two statements are unchanged. That is, if z is a root, then so is $\bar{z}$.

To summarize what we've learned so far, the roots of $f(z) = (1 + 2z)e^z - z = 0$ are infinite in number, are all complex with nonzero real parts, and each is one-half of a conjugate pair. But there still remains much more information to be mined. For example, I will next demonstrate for you that $x < 0$ for every root. Since we know $x = 0$ is impossible, then either $x > 0$ or $x < 0$. To set things up, first observe that, for z a root,

$$e^{-z} = 2 + \frac{1}{z}$$

and that the right-hand side is always defined since $z \neq 0$ (because $x \neq 0$) for every root.

For notational convenience, let us write $T = e^{-z}$ and $t = 2 + 1/z$. Then, $T = t$ and so, of course, $|T| = |t|$. Now,

$$|T| = |e^{-(x+iy)}| = |e^{-x}| \, |e^{-iy}| = |e^{-x}| = e^{-x}.$$

Suppose now that $x > 0$. Then $|T| < 1$. What I will do next is to show you that this leads immediately to a contradiction. We have

$$t = 2 + \frac{1}{x + iy} = 2 + \frac{x}{x^2 + y^2} - i \, \frac{y}{x^2 + y^2}.$$

Thus,

$$|t|^2 = \left(2 + \frac{x}{x^2 + y^2}\right)^2 + \left(\frac{y}{x^2 + y^2}\right)^2,$$

which, for $x > 0$, says $|t|^2 > 4$, or $|t| \geq 2$. That is, under the assumption of $x > 0$, we have concluded that $|T| < 1$ and $|t| > 2$, which violates the condition of $|T| = |t|$. Therefore, we conclude that the *premise* of $x > 0$ must be false. So it must be true that $x < 0$, i.e., all the roots must lie strictly to the left of the imaginary axis.

The same sort of reasoning will tell us, however, that the roots cannot be just anywhere in the left half of the complex plane. There are further restrictions on the roots. For example, let us write the roots of $f(z) = 0$ as $z = -x + iy$, with $x > 0$. Then $|T| = e^x$. Also,

$$t = 2 + \frac{1}{-x + iy} = 2 - \frac{x}{x^2 + y^2} - i\frac{y}{x^2 + y^2}$$

and so

$$|t|^2 = \left(2 - \frac{x}{x^2 + y^2}\right)^2 + \left(\frac{y}{x^2 + y^2}\right)^2.$$

Or, upon expanding,

$$|t|^2 = 4 - 4\frac{x}{(x^2 + y^2)^2} + \frac{1}{(x^2 + y^2)^2}.$$

Thus, as the second term is always positive (since we are assuming that $x > 0$), we have

$$|t|^2 < 4 + \frac{1}{(x^2 + y^2)^2} < 4 + \frac{1}{x^4},$$

and so, if $x > 1$, it would be true that

$$|t| < \sqrt{4 + \frac{1}{x^4}} < \sqrt{5} < 2.5 < e\, (= 2.718\ldots).$$

But, as shown before, $|T| = e^x$ and so, for $x > 1$, we have $|T| > e$. Again we have a violation of $|T| = |t|$, and so the premise of $x > 1$ must be false. Thus, the roots $z = x + iy$ must all be such that $-1 < x < 0$, i.e., all of the infinity of roots of $f(z) = 0$ lie in the vertical strip defined by $-1 < x < 0$, parallel to the imaginary axis.

There is, in fact, a lot more yet that can be said about the roots of $f(z) = 0$, and Professor Pierpont says it all in his paper, but this is as far as I'll go with the discussion. You will have to read his paper for the rest of it. But don't you think it absolutely astonishing how much information on the nature of the roots Professor Pierpont cleverly teased out of his equation, without ever coming anywhere near to actually calculating anything?

$(\sqrt{-1})^{(\sqrt{-1})}$ to 135 Decimal Places, and How It Was Computed

QUESTION: Is $(-1)^{\sqrt{-163}}$ an integer? Don't be too hasty in rejecting such a possibility—after all, as shown in section 3.3, the perhaps even more complicated-looking expression $e^{\pi\sqrt{-1}}$ is an integer. The answer is given at the end of note 4 for this appendix, but don't look there until you've read through what follows. If you do that, your appreciation of the surprising answer will be vastly greater than if you "peek."

It seems to be a common trait among curious, inventive minds to be fascinated by the calculation of certain special numbers. The cases of π and e in this respect are well known, particularly π, which has been calculated to many millions of decimal places. When Isaac Newton was developing his calculus, he used it to derive a convergent series for π which he then used to calculate the number π to sixteen places. He later wrote, as the explanation for why he did this, "I am ashamed to tell you to how many figures I carried these computations, having no other business at that time."[1] The great Gauss was attracted to equally detailed calculations as well, and after his death manuscripts were found in which he had calculated such numbers as $e^{\pi/2}$ and $2e^{-9\pi}$ to dozens of decimal places.

In 1921 an even longer calculation was performed that greatly extended Euler's wonderful calculation of $(\sqrt{-1})^{\sqrt{-1}}$, and the *method* of calculation is at least as interesting as the result.[2] In that year the Yale physicist Horace Scudder Uhler (1872–1956) published the value of i^i out to more than fifty decimal places. Before continuing with how he did this, I must tell you that Professor Uhler was no dilettante at number crunching. It was a life-long, serious obsession. In 1947, for example, he turned his attention to the largest decimal number that can be written using just three digits ($9^{(9^9)}$), which is known to have 369,693,100 digits. He calculated the logarithm of this monster to 250 decimal places. Why? He said he found such doings relaxing, and I think we should take him at his word.

Now, back to i^i. The obvious, brute force approach is to simply take a very long decimal value for π and use it to calculate $e^{-\pi/2}$ via the power series expansion for e^x—using a very extensive table of logarithms. However, to avoid any dependency on such decimal values and the risk of being done in by unknown errors that might be lurking in tabulated values, Uhler took a different, unexpected approach. At least, he *said* he did it "in order to avoid the use of any table of mathematical constants," but I suspect he did what he did just to have a bit of fun doing something that had never been done before.

What Uhler did was to define the function $f(x) = e^{\sin^{-1}(x)}$ and expand it into a power series. He did this because, for $x = \pm\frac{1}{2}$, he observed that $f(\pm\frac{1}{2}) = e^{\pm\pi/6}$. Then, cubing this, he had $e^{\pm\pi/2}$ (and so he had $i^i = e^{-\pi/2}$ without knowing π explicitly). He describes in his paper how he first determined that

$$f(x) = 1 + x + \frac{1}{2!}x^2 + \frac{2}{3!}x^3 + \frac{5}{4!}x^4 + \cdots$$

$$= \left(1 + \sum_{k=1}^{\infty} t_{2k+1}\right) + \left(x + \sum_{k=1}^{\infty} t_{2k+2}\right), \ |x| < 1,$$

where

$$t_{2k+1} = \frac{(0^2+1)(2^2+1)(4^2+1)\cdots[(2k-1)^2+1]x^{2k}}{(2k)!},$$

$$t_{2k+2} = \frac{(1^2+1)(3^2+1)(5^2+1)\cdots[(2k-1)^2+1]x^{2k+1}}{(2k+1)!}.$$

He used these general formulas to calculate eighty-five terms, each, for the even and odd powers of x, with each term rounded to fifty-four decimal places (5×10^{-55} was rounded up to 1×10^{-54}). Uhler concluded that this process gave him fifty-two correct decimal places of $e^{\pm\pi/6}$. His confidence in the calculations was enhanced when he separately calculated $f(\frac{1}{2}) = e^{\pi/6}$ and $f(-\frac{1}{2}) = e^{-\pi/6}$ and then multiplied the two results together—to fifty-two decimal places the result was exactly unity.

Years later, however, Uhler expressed some concern over the "prohibitively slow convergence" of his power series,[3] at least as far as it being of use in calculating additional digits beyond the fiftieth place. So, for a new, even more heroic computation, Uhler changed his approach to the "obvious" one I mentioned earlier, that of using 137-place tables of logarithms to calculate e^x from its power series expansion. Indeed, he used two different tables and checked the results from each table against the other with the goal of catching printing errors in the tables.

From one table he calculated both $e^{\pi/2}$ and $e^{-\pi/2}$ and checked the results by computing their product—he got 150 consecutive 9s after the decimal point. Then, with the other table, he calculated $e^{-\pi/192}$ and by successive squaring determined $e^{-\pi/6}$ and $e^{-\pi/3}$. Finally, multiplying these last two numbers together gave him, once again, a value for $e^{-\pi/2}$. This he compared to the value he had obtained from the other table. He found the two in perfect agreement out to the 139th place, with the first discrepancy in the 140th place. Because of this Uhler wrote that he "felt justified in guaranteeing the accuracy of 136" places. My personal opinion is that Professor Uhler is correct beyond any reasonable doubt in this assertion.[4]

Can you *imagine* the effort involved in doing all this? I quote Uhler: "The most error-sensitive part of the work consisted as usual in multiplying together mentally [remember, in 1947 Bill Gates, Microsoft, and a personal computer in every room were a long way in the future] the binomial factors as obtained from the logarithms. Hence, this set of operations was performed three times." The only explanation I have for this is that, for Uhler, playing with numbers was like a preschool kid playing with mud.

With no further discussion then, here is the value of i^i to 135 decimal places, where the final digit of 3 is the result of simply truncating the 136th digit of 9. Euler himself would surely have been impressed.

$$\left(\sqrt{-1}\right)^{\left(\sqrt{-1}\right)} = i^i = 0.20787 \ 95763 \ 50761 \ 90854 \ 69556 \ 19834 \ 97877 \ 00338$$
$$77841 \ 63176 \ 96080 \ 75135 \ 88305 \ 54198 \ 77285 \ 48213$$
$$97886 \ 00277 \ 86542 \ 60353 \ 40521 \ 77330 \ 72350 \ 21808$$
$$19061 \ 97303 \ 74663.$$

Notes

1. There was a science fiction hook in me for just -1, too, all by itself, even before taking its square root. Before *Popular Electronics* entered the picture I was a devoted fan of the 1950s weekly radio drama "X Minus One." Once a week, at 9:30 p.m., I listened to the opening of that program, to the announcer saying, "From the far horizons of the unknown come transcribed tales of new dimensions in time and space. These are stories of the future, adventures in which you'll live in a million could-be years and a thousand maybe worlds. The National Broadcasting Company, in cooperation with *Galaxy Science Fiction* magazine, presents . . . " And then came spine-chilling words, words reverberating in my darkened bedroom (when I was supposed to be asleep), "X x x x, minus, minus, minus, minus, one, one, one, one," followed by spooky music. Ah, those were the days!

INTRODUCTION

1. Richard J. Gillings, *Mathematics in the Time of the Pharaohs*, MIT Press 1972, pp. 246–47.

2. George Sarton, *A History of Science* (volume 1), Harvard University Press 1952, p. 39.

3. For interesting speculations on how the Egyptians might have reasoned, see Gillings (note 1), pp. 187–93, and B. L. van der Waerden, *Science Awakening*, P. Noordhoff 1954, pp. 34–35. This last book has a photograph of the frustum calculation portion of the MMP.

4. W. W. Beman, "A Chapter in the History of Mathematics," *Proceedings of the American Association for the Advancement of Science* 46 (1897):33–50.

5. The definitive work on Diophantus is still Sir Thomas L. Heath's *Diophantus of Alexandria: A Study in the History of Greek Algebra*, first published in 1885 and revised in 1910 (the revision is available today as a Dover reprint). This classic work contains the complete *Arithmetica* as Heath knew it, along with Heath's very extensive and helpful footnotes. In more recent times additional books have been discovered—see Jascques Sesiano, *Books IV to VII of Diophantus' Arithmetica*, Springer-Verlag 1982.

6. *The Ganita-Sara-Sangraha of Mahaviracarya* (with English translation and notes by M. Rangacarya), The Government Press, Madras 1912, p. 7 (of the translation).

CHAPTER ONE
THE PUZZLES OF IMAGINARY NUMBERS

1. This is not to say mathematicians didn't know what to make of, say, $7 - 5$. The distinction is between the use of the minus sign to denote the operation of subtraction

(which was understood), as opposed to denoting a number less than zero or nothing (which was mysterious). Thus, $5-7$ *would* have been a problem, since the answer of -2 would have had no physical significance to the mind of a mathematician at the start of the sixteenth century. Minus two would sometimes be called the *defect* of two, with the pejorative implication of that term perhaps revealing how uncomfortable negative numbers made mathematicians.

2. If you're wondering why I am ignoring the negative root, that's good. You *should* be wondering. The negative root is perfectly valid, but if you use it from this point on in the analysis you'll find that you'll get exactly the same answer as I will get with the positive root. Try it and see.

3. An informative and entertaining biography of Cardan's amazing life is the older but still recommended book by Oystein Ore, *Cardano, the Gambling Scholar,* Princeton 1953. Any man who is modern enough in intellect to solve the cubic, and still medieval enough to cast the horoscope for Christ, for which he was imprisoned in 1570 on the charge of heresy, is worth reading about.

4. An English translation of *Ars Magna,* by T. Richard Witmer, was published in 1968 by the MIT Press, and the following quote (p. 8) is the first of three mentions of Tartaglia by Cardan: "In our own days Scipione del Ferro of Bologna has solved the case of the cube and the first power equal to a constant, a very elegant and admirable accomplishment. Since this art surpasses all human subtlety and the perspicuity of mortal talent and is a truly celestial gift and a very clear test of the capacity of men's minds, whoever applies himself to it will believe that there is nothing that he cannot understand. In emulation of him, my friend Niccolo Tartaglia of Brescia, wanting not to be out-done, solved the same case when he got into a contest with his [Scipione's] pupil, Antonio Maria Fior, and, moved by my many entreaties, gave it to me." These are hardly the words of a man stealing the work of another. I should mention that there is some evidence that progress toward solving the cubic had been made no later than the 1390s in Florence, by at least two Italian mathematicians whose identities are no longer known. Indeed, del Ferro and Cardan may have been anticipated, although it is not at all clear that either knew of the earlier work. See R. Franci and L. Toti Rigatelli, "Towards a History of Algebra from Leonardo of Pisa to Luca Pacioli," *Janus* 72 (1985):17–82.

5. A modern student of mathematics, science, and engineering would find this claim obvious. That is, given any complex number, its conjugate is the number with all occurrences of $\sqrt{-1} = i$ changed to $-\sqrt{-1} = -i$.

6. Viète was the first to express π in terms of an infinite product rather than as a sum, in his famous formula (1593)

$$\frac{2}{\pi} = \cos\left(\frac{\pi}{4}\right) \cos\left(\frac{\pi}{8}\right) \cos\left(\frac{\pi}{16}\right) \cos\left(\frac{\pi}{32}\right) \cdots.$$

Later, in chapter 3, I will show you how this beautiful result, a special case of a more general trigonometric formula, can be derived from an elementary trigonometric identity that we can easily find with a knowledge of complex geometry. Viète's study of right triangles has been, in modern times, connected to the algebra of complex numbers, a connection he himself never made. I will discuss (in chapter 3) the known historical development of that algebra, but you can find what has been speculated in Stanislav Glushkov, "An Interpretation of Viète's 'Calculus of Triangles' As a Precursor of the Algebra of Complex Numbers," *Historia Mathematica* 4 (May 1977):127–36.

7. E. T. Bell, *The Development of Mathematics* (2nd edition), McGraw-Hill 1945, p. 149.

8. R. B. McClenon, "A Contribution of Leibniz to the History of Complex Numbers," *American Mathematical Monthly* 30 (November 1930):369–74. Huygens was just as puzzled as was Leibniz, as is shown by his reply to Leibniz: "One would never have believed that $\sqrt{1+\sqrt{-3}} + \sqrt{1-\sqrt{-3}} = \sqrt{6}$ and there is something hidden in this which is incomprehensible to us."

9. Gilbert Strang, *Introduction to Applied Mathematics,* Wellesley-Cambridge Press 1986, p. 330.

10. A nice English translation of this book by E. L. Sigler is available: *The Book of Squares,* Academic Press 1987. See also R. B. McClenon, "Leonardo of Pisa and His *Liber quadratorum,*" *American Mathematical Monthly* 26 (January 1919):1–8. Leonardo is better known today by the sobriquet "Fibonacci," which is most likely a contraction for "Filiorum Bonacci" (i.e., "of the family of Bonacci") or "Filius Bonacci" (i.e., "Bonacci's son"), phrases that appear in the titles of many of his works. It is, therefore, Leonardo Pisano after whom the famous Fibonacci sequence is named, i.e., the sequence 1, 1, 2, 3, 5, 8, 13, . . . , in which each number after the second is the sum of the previous two. This is usually written as the *recurrence* formula $u_{n+2} = u_{n+1} + u_n$ with $u_0 = u_1 = 1$. This sequence appears in Leonardo's *Liber abaci* (1202) and it is a special case of the more general $u_{n+2} = pu_{n+1} + qu_n$ with p and q arbitrary constants. In chapter 4 I will show you how complex numbers appear in the solution to this recurrence for certain values of p and q.

CHAPTER TWO
A FIRST TRY AT UNDERSTANDING THE GEOMETRY OF $\sqrt{-1}$

1. *La Geometrie* was not a book in its own right, but rather it was "just" the third illustrative appendix to Descartes' *Discourse on the Method of Reasoning and Seeking Truth in Science*. It was called "the greatest single step ever made in the progress of the exact sciences" by John Stuart Mill, and it *is* a masterpiece. It is well worth the time and effort, especially by high school geometry teachers, to read the excellent English translation by David E. Smith and Marcia L. Latham, *The Geometry of René Descartes,* Dover 1954.

2. Descartes himself offers no proof but Smith and Latham (note 1) sketch a possible approach. Their suggestion is, I think, more involved than necessary, and there is a much easier solution. Hint: look at the triangle *NQR* in figure 2.4 and use the Pythagorean theorem twice to find *QR*. Then $MQ = \frac{1}{2}a - \frac{1}{2}QR$ and $MR = MQ + QR$.

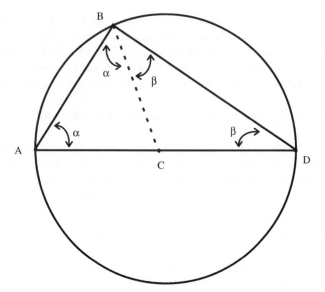

Figure 2.11. An inscribed right angle subtends a semicircle.

3. When the expression for T is real, then the smaller of the two possible values, the one given by using the minus sign, represents the time at which the man first catches the bus. But if the man does not hop onto the bus, but rather keeps on running, he will move out in front of the bus. You can easily verify that the man is indeed running faster than the bus is moving—after all, that's why he caught up to it. But the bus is accelerating and so eventually, at the second larger value of T, the bus catches up to the running man and again, for the second time, the condition $x_m = x_b$ is satisfied.

4. For much more on this particular problem see Nathan Altshiller Court, "Imaginary Elements in Pure Geometry—What They Are and What They Are Not," *Scripta Mathematica* 17 (1951):55–64 and 190–201. For more on the topic, in general, see J. L. S. Hatton, *The Theory of the Imaginary in Geometry, Together with the Trigonometry of the Imaginary,* Cambridge University Press 1920.

5. Florian Cajori, "Historical Note on the Graphic Representation of Imaginaries Before the Time of Wessel," *American Mathematical Monthly* 19 (October–November 1912):167–71.

6. This beautiful result is a special case of an even more general theorem concerning angles inscribed within circles, but since I'll use only this special case again before you are through this chapter, let me show you a quick proof of just the special case. With reference to figure 2.11, we have $AC = BC = CD$ as all three line segments are radii of the same circle. Since $AC = BC$, then the triangle ABC is isosceles, i.e., the angles $\angle$ BAC and $\angle$ ABC are equal (to α). And since $BC = CD$, then the triangle BCD is also isosceles, i.e., the angles $\angle$ CBD and $\angle$ CDB are equal (to β). Thus, $\alpha + \beta + (\alpha + \beta)$

$= 180° = 2(\alpha + \beta)$, or $\alpha + \beta = 90°$. That is, an angle inscribed on the circumference of a circle, which subtends a diameter of that circle, is a right angle. We can now use this result to quickly establish Descartes' construction of the square root of a positive length. Referring back to figure 2.1, look at the two right triangles FIG and IGH, which share the common side IG. We now know that the angle $\measuredangle$ FIH is 90°, and from this it immediately follows that the triangles FIG and IGH are similar. In particular,

$$\frac{GH}{IG} = \frac{IG}{FG} = \frac{IG}{1}$$

and so $IG = \sqrt{GH}$. The construction and this proof of it were known to the Greeks.

7. Julian Lowell Coolidge, *The Geometry of the Complex Domain,* Oxford 1924, p. 14.

CHAPTER THREE
THE PUZZLES START TO CLEAR

1. It is traditional to call Wessel a Norwegian, but in fact when he was born Norway was actually part of Denmark, and he spent most of his life in Denmark, dying in Copenhagen.

2. Wessel's paper was unearthed in 1895 by an antiquarian, and its significance recognized by the Danish mathematician Sophus Christian Juel (1855–1935). For more on this discovery, see Viggo Brun, "Caspar Wessel et l'introduction géométrique des numbres complexes," *Revue d'Histoire des Sciences et de Leurs Applications* 12 (1959):19–24.

3. E. T. Bell, *Men of Mathematics.* Simon and Schuster 1986, p. 234. A joke on this point is the following spoof of recorded telephone company messages that electrical engineers find amusing: "You have reached an imaginary number. If you require a real number, please rotate your telephone by 90°, and try again."

4. James Gleick, *Genius,* Pantheon 1992, p. 35.

5. See my book *The Science of Radio,* AIP Press 1995, pp. 173–75, for how such expressions occur in the theory of single-sideband radio.

6. The exchanges that appeared in the *Annales* were reproduced by Hoüel's book, and much of it was again reproduced in the English translation of Hoüel's reprint that was prepared by Dartmouth mathematics professor A. S. Hardy, *Imaginary Quantities,* D. Van Nostrand 1881.

7. Julian Lowell Coolidge, *The Geometry of the Complex Domain,* Oxford 1924, p. 24.

8. G. Windred, "History of the Theory of Imaginary and Complex Quantities," *The Mathematical Gazette* 14 (1929):533–41.

9. The mathematician was Sylvestre Francois Lacroix (1765–1843), who was famous for his enormously influential textbooks.

10. It is briefly discussed by Coolidge and Windred (notes 7 and 8), with the general assessment being that Mourey's book (*La Vraie Théorie des Quantités Négatives et des Prétendues Imaginaires*) was a work of quality.

11. See Bell's biographical essay on Hamilton, "An Irish Tragedy," *Men of Mathematics* (note 3), pp. 340–61.

12. John O'Neill, "Formalism, Hamilton and Complex Numbers," *Studies in History and Philosophy of Science* 17 (September 1986):351–72.

13. See my *Oliver Heaviside: Sage in Solitude,* IEEE Press 1988, pp. 187–215, for more on Hamilton and his quaternions. More interesting (in my opinion) than Hamilton's couples are the quadruples of numbers that form 2×2 matrices. For readers who know how matrices work, notice that

$$\begin{pmatrix} -1 & 0 \\ 0 & -1 \end{pmatrix}$$

plays the role of minus one, because multiplying it by any 2×2 matrix results in the negative of that matrix. Then simply notice that

$$\begin{pmatrix} 0 & 1 \\ -1 & 0 \end{pmatrix}\begin{pmatrix} 0 & 1 \\ -1 & 0 \end{pmatrix} = \begin{pmatrix} -1 & 0 \\ 0 & -1 \end{pmatrix},$$

and so

$$\begin{pmatrix} 0 & 1 \\ -1 & 0 \end{pmatrix}$$

plays the role of $\sqrt{-1}$ in 2×2 matrix theory. Such matrices, for example, are at the heart of quantum theory.

14. Quoted from Umberto Bottazzini, *The Higher Calculus: A History of Real and Complex Analysis from Euler to Weierstrass,* Springer-Verlag 1986, p. 96.

15. John Stillwell, *Mathematics and Its History,* Springer-Verlag 1991, p. 188.

16. A. R. Forsyth, "Old *Tripos* Days at Cambridge," *The Mathematical Gazette* 19 (1935):162–79.

CHAPTER FOUR
USING COMPLEX NUMBERS

1. The numerical examples used in this section are taken from Juan E. Sornito, "Vector Representation of Multiplication and Division of Complex Numbers," *Mathematics Teacher* 47 (May 1954):320–22, 382.

2. See, for example, E. F. Krause, *Taxicab Geometry,* Dover 1986, where the so-called city-block distance function $ds = |dx| + |dy|$ is explored in great detail. The geometric interpretation of special relativity is actually due to the German mathematician Hermann Minkowski (1864–1909), who was at one time Einstein's teacher. The theory, however, is all Einstein's.

3. See, for example, my book *Time Machines: Time Travel in Physics, Metaphysics, and Science Fiction,* AIP Press 1993, pp. 287–303.

4. This is actually only true for what is called the *flat* spacetime of special relativity. *Curved* spacetimes, such as occur in the general theory of relativity (Einstein's theory of gravity), have more complicated metrics. You can find more on this in my *Time Machines* (note 3), p. 314.

5. For why Einstein made this assumption see my *Time Machines* (note 3), pp. 291 and 302–3.

6. In 1889 the Russian mathematician Sophie Kowalevski (1850–91) used *complex* time to study the mechanics of a rotating mass. Her work is described in Michèle Audin, *Spinning Tops,* Cambridge University Press 1996.

7. Paul R. Heyl, "The Skeptical Physicist," *Scientific Monthly* 46 (March 1938):225–29.

CHAPTER FIVE
MORE USES OF COMPLEX NUMBERS

1. See also my *Time Machines: Time Travel in Physics, Metaphysics, and Science Fiction,* AIP Press 1993, pp. 341–52.

2. Edward Kasner, "The Ratio of the Arc to the Chord of an Analytic Curve Need Not Be Unity," *Bulletin of the American Mathematical Society* 20 (July 1914):524–31.

3. A brief but elegant, geometry-only derivation of the inverse square law is in George Gamow's *Gravity,* Doubleday 1962. See also S. K. Stein, "Exactly How Did Newton Deal with His Planets?" *Mathematical Intelligencer* 18 (Spring 1996):6–11.

4. See David L. and Judith R. Goodstein's *Feynman's Lost Lecture,* W. W. Norton 1996. The book comes with a compact disk recording, and so you can actually listen to Feynman give his lecture to the Caltech freshman class on March 13, 1964. At the end of his lecture Feynman makes the important point that exactly the same physics is at the heart of Ernest Rutherford's classic 1910 experiment in scattering particles with positive electric charge off the nuclei of atoms which also have positive charge. As will be discussed later in this chapter, the interaction force is *repulsive* between charges of the same sign, not attractive as is the gravitational force between the Sun and the orbiting planets. But this is, mathematically, simply a trivial change of algebraic sign in the force equation.

5. The presentation in this section was inspired by Donald G. Saari, "A Visit to the Newtonian *N*-Body Problem via Elementary Complex Variables," *American Mathematical Monthly* 97 (February 1990):105–19.

6. William J. Kaufmann III, *Universe* (2nd edition), W. H. Freeman 1988, p. 56.

7. See my book *The Science of Radio,* AIP Press 1995, for the details of how that came to pass.

CHAPTER SIX
WIZARD MATHEMATICS

1. Wroński's claim was not unique in the history of weird mathematics. What in some way is an even more bizarre assertion was made at the end of the nineteenth

century, in what has since become one of the most prestigious mathematics journals in the world. In a biographical essay on Colonel James W. Nicholson (1844–1917), who was President and Professor of Mathematics at Louisiana State University and Agricultural and Mechanical College, the claim is made that among *his* many discoveries are the identities

$$\cos(\phi) = \frac{(-1)^{\phi/\pi} + (-1)^{-\phi/\pi}}{2},$$

$$\sin(\phi) = \frac{(-1)^{\phi/\pi} - (-1)^{-\phi/\pi}}{2\sqrt{-1}}.$$

This is absurd, of course, as the identities are simply thinly disguised versions of Euler's identity, i.e., just write $-1 = e^{i\pi}$. I have no explanation for how such a silly claim could be printed; at first I was convinced it had to be a spoof, but I soon discovered Nicholson was a real person. If you want to read the entire essay for yourself (in which additional absurd claims can be found) see *American Mathematical Monthly* 1 (June 1894):183–87.

2. How he did this, and much more, can be found in Raymond Ayoub, "Euler and the Zeta Function," *American Mathematical Monthly* 81 (December 1974):1067–86. See also Ronald Calinger, "Leonhard Euler: The First St. Petersburg Years (1727–1741)," *Historia Mathematica* 23 (May 1996):121–66.

3. Much of my discussion on i^i in this chapter is based on R. C. Archibald, "Historical Notes on the Relation $e^{-(\pi/2)} = i^i$," *American Mathematical Monthly* 21 (March 1921):116–21.

4. For a full English translation of "Logometria" see Ronald Gowing, *Roger Cotes—Natural Philosopher,* Cambridge University Press 1983.

5. R. C. Archibald, "Euler Integrals and Euler's Spiral—Sometimes Called Fresnel Integrals and the Clothoide or Cornu's Spiral," *American Mathematical Monthly* 25 (June 1918):276–82.

6. Historical comments about, and citations to, the work of Cauchy and Poisson related to these integrals can be found in Horace Lamb, "On Deep-Water Waves," *Proceedings of the London Mathematical Society* 2 (November 10, 1904):371–400.

7. In two letters, dated October 13, 1729 and January 8, 1730, to one of his frequent correspondents, the German Christian Goldbach in Moscow. The name and symbol of *gamma* is, however, due to Legendre, who introduced the modern terminology in 1808. For a tutorial on this part of Euler's work, see Philip J. Davis, "Leonhard Euler's Integral: A Historical Profile of the Gamma Function," *American Mathematical Monthly* 66 (December 1959):849–69.

CHAPTER SEVEN

THE NINETEENTH CENTURY, CAUCHY, AND THE BEGINNING OF COMPLEX FUNCTION THEORY

1. Much of my commentary on Cauchy's work is based on H. J. Ettlinger, "Cauchy's Paper of 1814 on Definite Integrals," *Annals of Mathematics* 23 (1921–22):255–

70, and on Philip E. B. Jourdain, "The Theory of Functions with Cauchy and Gauss," *Bibliotheca Mathematica* 6 (1905):190–207. Neither Ettlinger nor I follow Cauchy's original presentation or notation slavishly (Jourdain does, largely), although the *ideas* he developed will be pointed out. Very interesting reading, too, is the second half of volume 2 of the series *Mathematics of the 19th Century,* on the history of analytic function theory. Edited by A. N. Kolmogorov and A. P. Yushkevich, it was translated from the Russian by Roger Cooke and published by Birkhäuser Verlag in 1996. I also found helpful the extensive historical commentary in Morris Kline, *Mathematical Thought from Ancient to Modern Times,* Oxford University Press 1972, pp. 626–70. Some of Kline's comments concerning the original debates over the 1814 memoir were clarified for me in I. Grattan-Guinness, *The Development of the Foundations of Mathematical Analysis from Euler to Riemann,* MIT Press 1979, pp. 24–45. Finally, see Frank Smithies, *Cauchy and the Creation of Complex Function Theory,* Cambridge University Press 1997.

2. Cauchy didn't use this term, but the concept of analyticity is contained in his memoir. And, as Ettlinger (note 1) observes, "although geometrical representation is now an essential feature of every presentation of the theory of functions, Cauchy used neither figures nor geometrical language." Following Ettlinger and all modern texts I use both aids in this chapter a *lot.*

3. This very point was the central concept in a curious short science fiction story by Stanley G. Weinbaum. Originally titled "Real and Imaginary" it appeared, after Weinbaum's early death, in the December 1936 issue of the pulp magazine *Thrilling Wonder Stories* under the title "The Brink of Infinity." It is reprinted in *A Martian Odyssey and Other Science Fiction Tales,* Hyperion Press 1974. The plot is a simple one. A chemical expert is horribly mangled when an experiment goes wrong, and he blames the mathematician who did the preliminary calculations. In revenge on mathematicians, in general, he lures another mathematician to his home and, holding him at gun point, presents him with the following problem: the madman is thinking of a "numerical expression" and the mathematician must deduce what it is by asking no more than ten questions. If he can't, he'll be shot. The first question is whether the expression is real or imaginary, and the answer is "either"! The mathematician thinks that means the answer must be zero, and is shocked to be told he is wrong. Another question is, "What is the expression equal to?" and the answer is "anything." And that's the point here, that eventually the mathematician figures out that the answer is "$\infty - \infty$," although I doubt this is properly called a *numerical* expression.

4. The fact that we had to use (as did Cauchy) the value of $\int_0^\infty e^{-x^2}dx$ (known from other means, such as was done in section 6.12) to complete the calculation of $\int_0^\infty e^{-x^2}\cos(2bx)dx$ illustrates the fact that Cauchy's method isn't without its drawbacks. In fact, any integral that can be done by Cauchy's method can be done by other means. See, for example, Robert Weinstock, "Elementary Evaluations of $\int_0^\infty e^{-x^2}dx$, $\int_0^\infty \cos x^2 dx$, and $\int_0^\infty \sin x^2 dx$," *American Mathematical Monthly* 97 (January 1990):39–42. However, if Cauchy's method does work then it is almost always the easiest way to evaluate an integral. On the flip side, for many years it was thought that $\int_0^\infty e^{-x^2}dx$ *couldn't* be done with Cauchy's method, but for a pretty derivation of this integral

using Cauchy's theory, see Reinhold Remmert, *Theory of Complex Functions,* Springer-Verlag 1991, pp. 413–4. More on the history of that derivation can be found in Dragoslav S. Mitrinović and Jovan D. Keckić, *The Cauchy Method of Residues,* D. Reidel 1984, pp. 158–66.

5. See volume 4 of Stokes' *Mathematical and Physical Papers,* Cambridge University Press 1904, pp. 77–109.

6. Jesper Lützen, *Joseph Liouville 1809–1882: Master of Pure and Applied Mathematics,* Springer-Verlag 1990, p. 586. See also the chapter on integral theorems by J. J. Cross in *Wranglers and Physicists: Studies on Cambridge Mathematical Physics in the Nineteenth Century* (P. M. Harman, editor), Manchester University Press 1985.

7. I. Grattan-Guinness, "Why Did George Green Write His Essay of 1828 on Electricity and Magnetism?" *American Mathematical Monthly* 102 (May 1995):387–96.

8. The Polish-born mathematician Mark Kac tells the following wonderfully funny story in his autobiography *Enigmas of Chance,* Harper & Row 1985, p. 126. Once, when serving on a doctoral examination committee at Cornell, Kac asked the candidate some mathematical questions. As Kac tells it, "He was not terribly good—in mathematics at least. After he had failed to answer a couple of questions, I asked him a really simple one, which was to describe the behavior of the function $1/z$ in the complex plane. 'The function is analytic, sir, in the whole plane except at $z = 0$, where it has a singularity,' he answered and it was perfectly correct. 'What is the singularity called?' I continued. The student stopped in his tracks. 'Look at me,' I said. 'What am I?' His face lit up. 'A simple Pole, sir,' which is in fact the correct answer." What a nice person Kac must have been!

9. Please don't think I am being disrespectful by calling Cauchy's solution "awkward." Newer, better ways of solving problems are welcomed in mathematics and, indeed, are *hoped for.* The method I use in the text used to be a common textbook approach, but even it has been improved upon. The answer derived in the text is actually valid even for non-integer values of m and n, but the method I use does impose that restriction. You can find more discussion on this important integral, and its more general evaluation, in Orin J. Farrell and Bertram Ross, "Note on Evaluating Certain Real Integrals by Cauchy's Residue Theorem," *American Mathematical Monthly* 68 (February 1968):151–52.

10. Such a book is the one by A. David Wunsch, *Complex Variables with Applications* (2nd edition), Addison-Wesley 1994. Wunsch, a professor of electrical engineering at the Lowell campus of the University of Massachusetts, has done an absolutely first-rate job of presenting the details of complex function theory, including conformal mapping, Laurent series, and system stability. He writes in the language of the engineer, while at the same time giving up none of the mathematical integrity of the subject. For the more mathematically inclined who also have an interest in keeping one foot in physical reality, I highly recommend Tristan Needham, *Visual Complex Analysis,* Oxford University Press 1997.

11. For a discussion of the details of doing this integral, i.e., of evaluating

$$\int_0^\infty \frac{x^{s-1}}{e^x - 1}\, dx$$

in the complex plane, see E. C. Titchmarsh (revised by D. R. Heath-Brown), *The Theory of the Riemann Zeta-Function,* Oxford University Press 1986, pp. 18–20.

12. This story is reprinted in *The Early Asimov,* Doubleday 1972.

13. Marilyn vos Savant, *The World's Most Famous Math Problem (the Proof of Fermat's Last Theorem and Other Mathematical Mysteries),* St. Martin's Press 1993, p. 61. This book, which the author happily tells us she wrote in mere weeks, is full of other similarly uninformed statements.

14. A reproduction of this holographic letter is in *Scripta Mathematica* 1 (1932):88–90.

Appendix A
The Fundamental Theorem of Algebra

1. For more on all this, see John Stillwell, *Mathematics and Its History,* Springer-Verlag 1989, pp. 195–200.

2. An early, little-known dispute over the number of complex roots to a polynomial equation took place between the Scottish mathematician Colin MacLaurin (1698–1746) and his obscure countryman George Campbell (?–1766). A detailed discussion of this unhappy business is given by Stella Mills, "The Controversy Between Colin MacLaurin and George Campbell Over Complex Roots, 1728–1729," *Archive for History of Exact Sciences* 28 (1983):149–64. For even more on the state of knowledge concerning the roots of equations at that time see Robin Rider Hamburg, "The Theory of Equations in the 18th Century: The Work of Joseph Lagrange," *Archive for History of Exact Sciences* 16 (1976):17–36.

3. For example, solve $ix^2 - 2x + 1 = 0$ and show that $x = -i \pm \sqrt{-1 + i}$. Then show that $x = -i + \sqrt{-1 + i}$ and $x = -i - \sqrt{-1 + i}$ are *not* a conjugate pair. In fact, the conjugate of $x = -i + \sqrt{-1 + i}$ is $i + \sqrt{-1 - i} \neq -i - \sqrt{-1 + i}$.

4. E. T. Bell, *The Development of Mathematics* (2nd edition), McGraw-Hill 1945, p. 176.

Appendix B
The Complex Roots of a Transcendental Equation

1. James Pierpont, "On the Complex Roots of a Transcendental Equation Occurring in the Electron Theory," *Annals of Mathematics* 30 (1929):81–91. The inspiration for Professor Pierpont was G. A. Shott, "The Theory of the Linear Oscillator and Its Bearing on the Electron Theory," *Philosophical Magazine* 3 (April 1927):739–52.

Appendix C
$(\sqrt{-1}^{(\sqrt{-1})})$ to 135 Decimal Places and How It Was Computed

1. Newton (1642–1727) derived his fast-convergent series for π during 1665–66, when he fled plague-ridden London to return home to the family farm at Woolsthorpe in Lincolnshire. Thus, his series predates the one due to Leibniz-Gregory (discussed in

chapter 6), but it did not appear in print until many years later, in the 1736 posthumous publication of an English translation (*The Method of Fluxions and Infinite Series*) of his original Latin manuscript. It was during this same period in Woolsthorpe that Newton discovered but didn't publish the power series expansion for $\ln(1 + x)$ *before* Mercator (see the discussion in section 6.3). This is the series used by Schellbach to calculate π from $\sqrt{-1}$—again, as discussed in chapter 6—and which Newton used to calculate various "interesting" numbers out to as many as sixty-eight decimal places.

2. H. S. Uhler, "On the Numerical Value of i^i," *American Mathematical Monthly* 21 (March 1921):114–16.

3. Horace S. Uhler, "Special Values of $e^{K\pi}$, COSH($K\pi$) and SINH($K\pi$) to 136 Figures," *Proceedings of the National Academy of Sciences* 33 (February 1947):34–41.

4. With the development of powerful mathematical programming languages such as MATLAB (see figure 7.8), the calculation of i^i can be done in a flash with a digit count far beyond Uhler's. For example, the single line of code

$$\text{vpa}('i \wedge i', 140)$$

invokes the variable-precision arithmetic command which tells MATLAB to print i^i to 140 digits. My little lap-top did that in less than a second; the displayed answer was *exactly* Uhler's. And now you can see how to calculate the answer to the question that opens this appendix. To begin, notice that there are two values for the expression, i.e., since $-1 = e^{\pm i\pi}$ and since $\sqrt{-163} = \pm i\sqrt{163}$, then $(-1)^{\sqrt{-163}} = e^{\pm\pi\sqrt{163}}$. We can immediately reject $e^{-\pi\sqrt{163}}$ as an integer, because it is both greater than zero and less than one. But what of $e^{\pi\sqrt{163}}$? It is obviously pretty big, but is it an integer? Writing the MATLAB command

$$\text{vpa}('\exp(\text{pi} * \text{sqrt}(163))', 37)$$

gives the value of

$$(-1)^{\sqrt{-163}} = 262,537,412,640,768,743.99999999999925007,$$

and so the answer is *no,* the expression is not an integer. It is, in fact, transcendental. But it misses being an integer by oh so very, very little! For a theoretical explanation of what is behind this result, see Philip J. Davis, "Are There Coincidences in Mathematics?" *American Mathematical Monthly* 88 (May 1981):311–20.

Name Index

Abd er-Rassul, Ahmed and Mohammed, 3
Abel, Niels, 190
Airy, George, 83
Ampere, Andre-Marie, 126
Apéry, Roger, 149
Arago, Dominique, 142
Argand, Jean-Robert, 73–74, 76–78, 227
Aristarchus, 111
Aristotle, 111
Asimov, Isaac, 223
Auden, Wystan Hugh, 13–14, 224

Bannister, Roger, 15
Bell, Eric Temple, 78, 229
Beman, Wooster Woodruff, 4
Benedict XIV (Pope), 158
Bernoulli, Daniel, 142
Bernoulli, John, 91, 142–43, 145, 148, 157–60, 162, 216
Bernoulli, Nicolas, 142
Bombelli, Rafael, 18–19, 22–25, 31, 58
Boole, George, 83
Brahe, Tycho, 113
Buée, Adrien-Quentin, 75–77

Campbell, George, 249n
Cardan. See Cardano, Girolamo
Cardano, Girolamo ("Cardan"), 16–19, 24, 74, 224, 240 (n.3, n.4)
Carnot, Lazare, 75–76, 189
Carnot, Sadi, 75
Catherine I (Tsarina of Russia), 142
Catherine the Great (Tsarina of Russia), 143
Cauchy, Augustin-Louis, 53, 73, 120, 175, 179, 187–90, 193, 195, 197, 203, 208, 213–14, 221, 225, 246n, 247n
Cavendish, Henry, 115
Charles I (King of England), 41
Charles X (King of France), 190
Christ, 240
Christina (Queen of Sweden), 34
Christoffel, Elwin Bruno, 222

Clausen, Thomas, 79
Collins, John, 42
Copernicus, Nicolaus, 112–13
Cotes, Roger, 86, 91, 162–63, 165–66
Coulomb, Charles, 126
Cromwell, Oliver, 41

d'Alembert, Jean le Rond, 78, 193, 227–28
D'Arsonval, Arsène, 131
de Broglie, Louis, 79
del Ferro, Scipione, 8–13, 14–15, 240n
De Moivre, Abraham, 56–57, 162, 178
DeMorgan, Augustus, xvi, 82
Descartes, René, 6, 31–32, 34–37, 40, 43, 46, 227, 241n, 243n
Diderot, Denis, vii
Diophantes, 4–8, 26, 239n
Dirichlet, Gustav-Peter Lejeune, 182, 223
Disney, Walt, 141

Einstein, Albert, 97, 101, 103–05, 244n
Eratosthenes (of Cyrene), 151
Euclid, 4–5, 80, 150–51, 224
Euler, Leonhard, 10, 12, 14, 57, 78, 110, 142–47, 149–52, 157- 58, 166–67, 174–80, 182, 189, 201, 216–17, 225, 227, 235, 237, 246n

Fagnano, Giulio Carlo dei Toschi, 158–60, 218
Faraday, Michael, 127
Feynman, Richard, 67, 113–14, 245n
Fibonacci. See Pisano, Leonardo
Fior, Antonio Maria, 15, 240n
Fontana, Niccolo ("Tartaglia"), 15–16, 74, 240n
Fourier, Jean Baptiste Joseph, 222
Francais, Francois, 73
Francais, Jacques, 73–74, 77
Fredrick the Great (of Prussia), 143
Fresnel, Augustin Jean, 175

251

Subject Index

Acknowledgments

IT IS ALWAYS a pleasure for me to thank those who have helped me create a new book. At the University of New Hampshire the physicist Roy Torbert enthusiastically endorsed the idea of an electrical engineer writing a math-history book; since he is my dean, that support was most important! Also, Nan Collins typed it, Kim Riley created the line illustrations, Barbara Lerch helped me with the indexing, and my freshman General Education Honors Seminar class in the fall of 1996 read a very early draft with a sharp and critical collective eye. Professor Arturo Sangalli at Champlain Regional College (Quebec) read an early revision of the first draft, and his enthusiastic response was of greater value to me than he can know. At Loyola University, Caltech, and Harvey Mudd College, respectively, Professors Eli Maor, David Rutledge, and John Molinder read a nearly final revision and gave me a number of helpful comments. My sharp-eyed copy editor, Jennifer Slater, made what could easily have been a trying time a pleasant one. At Princeton University Press, my editor Trevor Lipscombe and his assistant Sam Elworthy, and production editor Karen Verde, transformed a pile of typescript—plastered with Jennifer's and my notes for changes, additions, and deletions—into a *book*. And finally, my wife of thirty-six years, Patricia Ann, patiently listened as I talked, mumbled, snarled, and sometimes sobbed to myself all through the many days and nights it took to do my writing. Not once did she tell me to please be quiet, and for that (but not *just* for that) do I love her dearly.

ABOUT THE AUTHOR

Paul J. Nahin is Professor of Electrical Engineering at the University of New Hampshire. He is the author of *Oliver Heaviside, Time Machines,* and *The Science of Radio,* and of two dozen short science fiction stories, published in *Omni, Analog,* and *Twilight Zone* magazines.